PETER GUTHRIE TAIT
WILLIAM JOHN STEELE

A TREATISE

ON THE

DYNAMICS OF A PARTICLE

WITH

NUMEROUS EXAMPLES

Elibron Classics
www.elibron.com

Elibron Classics series.

ISBN 1-4021-7104-8 (paperback)
ISBN 1-4021-2117-2 (hardcover)

This Elibron Classics Replica Edition is an unabridged facsimile of the edition published in 1856 by Macmillan and Co., Cambridge.

A TREATISE

ON THE

DYNAMICS OF A PARTICLE.

Cambridge:
Printed at the University Press,

For MACMILLAN & Co.

London: BELL AND DALDY.
Oxford: J. H. AND JAMES PARKER.
Edinburgh: EDMONSTON AND DOUGLAS.
Dublin: WILLIAM ROBERTSON.
Glasgow: JAMES MACLEHOSE.

A TREATISE

ON THE

DYNAMICS OF A PARTICLE

WITH

Numerous Examples.

BY

PETER GUTHRIE TAIT, M.A.

FELLOW OF ST PETER'S COLLEGE,
PROFESSOR OF MATHEMATICS IN THE QUEEN'S UNIVERSITY.

AND THE LATE

WILLIAM JOHN STEELE, B.A.

FELLOW OF ST PETER'S COLLEGE.

Cambridge:
MACMILLAN AND CO.
1856.

PREFACE.

In the present Treatise will be found all the ordinary propositions connected with the Dynamics of particles which can be conveniently deduced without the use of D'Alembert's Principle.

Its publication has been delayed by many unforeseen occurrences; more especially by the early and lamented death of Mr Steele, whose portion of the work was left uncompleted, and whose assistance in its final arrangement and revision would have been invaluable. The principal portions due to him are the greater part of Chapters III., V. and VIII. together with a few pages of Chapter I.

Considerable use has been made of Pratt's *Mechanical Philosophy:* indeed a large portion of Chapter XI. is reprinted verbatim from that work.

Throughout the book will be found a number of illustrative examples introduced in the text, and for the most part completely worked out; others with occasional solutions or hints to assist the student are appended to each Chapter. For by far the greater portion of these the Cambridge Senate-House and College Examination Papers have been applied to.

I am requested by the Publishers to state that a Treatise on the remaining departments of Dynamics, including the Motion of a Rigid System, &c. is in contemplation; and will appear as soon as circumstances will permit.

PETER G. TAIT.

QUEEN'S COLLEGE, BELFAST,
January, 1856.

ERRATUM.

Page 164, line 14, for g read mg.

CONTENTS.

INTRODUCTION.

The position of a particle in space is usually determined at any instant by its distances from any three fixed co-ordinate planes at right angles to each other, in the same manner as the position of a point is determined in Geometry of Three Dimensions.

Now it will be seen hereafter, that a material particle, uninfluenced by external forces, remains permanently at rest in space, or moves uniformly in a straight line. In the former case its co-ordinates are constant, in the latter their increments are proportional to the time and to each other. If they do not satisfy one or other of these conditions, there must be some cause for the change, and it is the object of the science of Dynamics to determine these changes of motion from the causes (supposed known) which produce them, and conversely, from the effects (supposed known) to discover the magnitude of the causes to which they are due.

Although the idea of some physical cause is necessary when we consider the effects produced on the motion of a particle, yet there are many properties of motion which may be arrived at without having recourse to anything but the abstract idea of motion itself. Such properties are to be regarded as purely geometrical, in contradistinction to others which depend on our experimental knowledge of the constitution of matter, and the idea of Force. These are called Kinematical, and form the subject of the first of the following Chapters.

DYNAMICS OF A PARTICLE.

CHAPTER I.

GEOMETRY OF A MOVING POINT'S MOTION.

1. VELOCITY is a term used to indicate the degree of quickness or slowness with which a body moves. Velocity is measured by the space passed over in a given time, and may be either uniform or variable.

Velocity is said to be uniform when the body passes over equal spaces in equal times, and is then measured by the actual space passed over in a given time. Let the unit be so chosen that a particle moving with an unit of velocity would pass over an unit of space in an unit of time, then a particle moving with v units of velocity would pass over v units of space in an unit of time, and consequently vt units of space in t units of time. Therefore

$$s = vt$$

where v denotes the velocity of a particle moving uniformly, and s the space described in the time t.

In this formula the units of space and time are perfectly arbitrary and may be chosen at convenience.

2. Velocity is said to be variable when the body in motion does not describe equal spaces in equal times. The velocity at any point of the path is then measured by the space which would have been described in an unit of time, if the particle after passing the point had moved on uniformly with the velocity which it had there. Let v be the velocity of a particle at the end of a time t measured from a fixed epoch, s the space described by it during that interval, and $s + \delta s$ the space described during a greater interval $t + \delta t$.

Suppose v_1 to be the greatest and v_2 the least velocity with which the particle is animated during the time δt, then $v_1\delta t$, $v_2\delta t$ would be the spaces which a particle would describe in that interval, moving uniformly with these velocities respectively. But the actual velocity of the particle is not greater than v_1 and not less than v_2, therefore the actual space described,

$$\delta s \text{ is not greater than } v_1\delta t \text{ and not less than } v_2\delta t,$$

$$\text{or } \frac{\delta s}{\delta t} \ldots\ldots\ldots\ldots\ldots\ldots\ldots\ldots \; v_1 \quad \ldots\ldots\ldots\ldots\ldots\ldots\ldots\ldots \; v_2,$$

however small δt may be. But as δt continually diminishes v_1 and v_2 tend continually to and ultimately become each equal to v. Therefore proceeding to the limit

$$\frac{ds}{dt} = v.$$

If v be negative in this expression, it indicates (TODHUNTER, *Diff. Calc.*, Art. 88.) that s diminishes as t increases; the positive case, which we have taken as the standard one, referring to that in which s and t increase together. If a velocity then in one direction be considered positive, that in the opposite must be considered negative, and consequently the sign of the velocity will indicate the direction of motion.

3. It will be easily seen that the above idea of velocity is equally applicable whether the particle be considered as moving in a straight or a curved line. In the latter case however the motion of the particle is constantly changing as to its direction, and it will be necessary at any instant to know the direction as well as the actual velocity of the particle. This is done by considering the velocities of the particle parallel to the three co-ordinate axes respectively. For if the co-ordinates of the particle be represented by x, y, z, the rates of increase of these, or the velocities parallel to them respectively, will by reasoning analogous to the above be

$$\frac{dx}{dt}, \quad \frac{dy}{dt}, \quad \frac{dz}{dt}.$$

Denoting by v the whole velocity of the particle, we have

$$v = \frac{ds}{dt} = \sqrt{\left\{\left(\frac{dx}{dt}\right)^2 + \left(\frac{dy}{dt}\right)^2 + \left(\frac{dz}{dt}\right)^2\right\}};$$

and if α, β, γ be the angles which the direction of motion makes with the axes,

$$\cos\alpha = \frac{dx}{ds} = \frac{\frac{dx}{dt}}{\frac{ds}{dt}};$$

$$\text{or } \frac{dx}{dt} = v\cos\alpha = v_x, \text{ suppose.}$$

Similarly,

$$\frac{dy}{dt} = v\cos\beta = v_y,$$

$$\frac{dz}{dt} = v\cos\gamma = v_z.$$

Hence, $\frac{dx}{dt}$, $\frac{dy}{dt}$, $\frac{dz}{dt}$ are the resolved parts of v parallel to the axes, and are therefore called the component velocities of the particle.

4. It follows from the above, that, if a particle be moving in any direction, we may suppose its velocity to be the resultant of three coexistent velocities in any three directions at right angles to each other.

The same is true if the axes are not at right angles, but such cases are of rare occurrence and of little importance.

Let v_x, v_y, v_z be the resolved parts of the velocity of particle moving with the velocity v, then the resolved part ot v along a line inclined at angles λ, μ, ν to the axes will be $v_x \cos\lambda + v_y \cos\mu + v_z \cos\nu$. For let α, β, γ be the angles which the direction of the particle's motion makes with the axes, θ the angle between this direction and the given line. Then the resolved part of v along that line

$$= v\cos\theta = v\{\cos\alpha\cos\lambda + \cos\beta\cos\mu + \cos\gamma\cos\nu\}$$

$$= v_x\cos\lambda + v_y\cos\mu + v_z\cos\nu.$$

5. *When a particle moves in a plane curve, to express its velocities at any instant along and perpendicular to the radius vector drawn from a fixed point in the plane of the curve.*

Let x, y be its rectangular, r, θ its polar co-ordinates, so that

$$x = r\cos\theta,\ \ y = r\sin\theta;$$

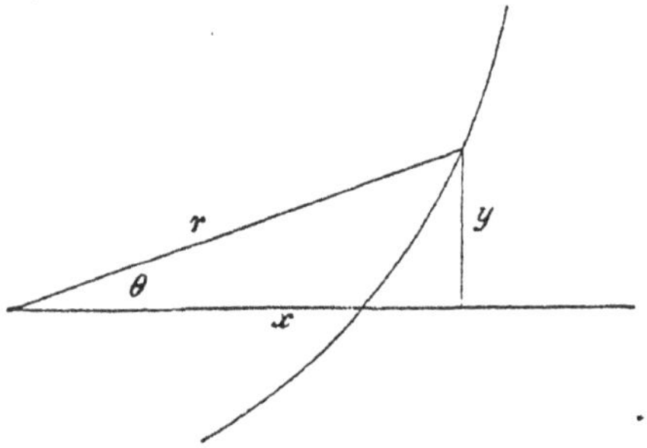

$$\therefore\ \frac{dx}{dt} = \frac{dr}{dt}\cos\theta - r\sin\theta\,\frac{d\theta}{dt},$$

$$\frac{dy}{dt} = \frac{dr}{dt}\sin\theta + r\cos\theta\,\frac{d\theta}{dt},$$

the velocities parallel to x and y. But by last Art. the velocity along the radius vector

$$= \frac{dy}{dt}\sin\theta + \frac{dx}{dt}\cos\theta = \frac{dr}{dt},$$

and the velocity perpendicular to it

$$= \frac{dy}{dt}\cos\theta - \frac{dx}{dt}\sin\theta = r\,\frac{d\theta}{dt}.$$

6. The velocity of a particle is said to be accelerated or retarded according as it increases or diminishes, but the word acceleration is generally used in both senses.

Acceleration of velocity may be either uniform or variable. It is said to be uniform when the particle receives equal increments of velocity in equal times, and is measured by the actual increase of velocity generated. Let the unit of acceleration be so taken that a particle under its action would receive an increment of an unit of velocity in an unit of time, then a particle under the influence of α units of acceleration would receive an increment of α units of velocity in an unit of time, and consequently αt units of acceleration in t units of time. If the particle then starts from rest

$$v = \alpha t,$$

where v denotes the velocity at the end of the interval t, and α the accelerating effect.

7. Acceleration of velocity is variable when the particle does not receive equal increments of velocity in equal increments of time. The acceleration at any instant is then measured by the increment of velocity, which would have been generated in an unit of time had the accelerating effect remained the same during that unit and equal to that at the given instant.

Let v be the velocity of the particle at the end of the time t, α the accelerating effect at that instant, $v + \delta v$ the velocity at the end of the time $t + \delta t$, and let α_1, α_2 be the greatest and least accelerations during the interval δt, then $\alpha_1 \delta t$, $\alpha_2 \delta t$ would be the increments of velocity in that interval, of a particle under those accelerating effects respectively. But the actual acceleration is not greater than α_1 and not less than α_2, therefore the actual increment of velocity

δv is not greater than $\alpha_1 \delta t$ and not less than $\alpha_2 \delta t$,

or $\frac{\delta v}{\delta t}$ .... α_1 .... α_2,

however small δt may be. But as δt continually diminishes α_1 and α_2 tend continually to and ultimately become each equal to α. Therefore proceeding to the limit

$$\frac{dv}{dt} = \alpha.$$

The positive sign given to α shews that v increases with t, while a negative sign would shew that v decreases with t, in other words a negative acceleration is a retardation.

Combining the above equation with

$$\frac{ds}{dt} = v,$$

we have

$$\frac{d^2s}{dt^2} = \alpha,$$

considering t as the independent variable.

If the particle be in motion along a curved line, the accelerations of the motions of increase of its co-ordinates are called the component accelerations of the particle's velocity parallel to the axes. If these be denoted by α_x, α_y, α_z, we shall have

$$\frac{d^2x}{dt^2} = \alpha_x, \quad \frac{d^2y}{dt^2} = \alpha_y, \quad \frac{d^2z}{dt^2} = \alpha_z.$$

8. The acceleration $\frac{d^2s}{dt^2}$ is not the complete resultant of $\frac{d^2x}{dt^2}$, $\frac{d^2y}{dt^2}$, $\frac{d^2z}{dt^2}$, as may easily be seen because its square does not equal the sum of the squares of those three accelerations, but it is only that part of the resultant which has any effect on the velocity; in short $\frac{d^2s}{dt^2}$ is the sum of the resolved parts of $\frac{d^2x}{dt^2}$, $\frac{d^2y}{dt^2}$, $\frac{d^2z}{dt^2}$ in the direction of motion, as the following identical equation shews:

$$\frac{d^2s}{dt^2} = \frac{dx}{ds}\frac{d^2x}{dt^2} + \frac{dy}{ds}\frac{d^2y}{dt^2} + \frac{dz}{ds}\frac{d^2z}{dt^2}.$$

The other part of the resultant is at right angles to this, and shews its effect in changing the direction of the motion of the particle. Its value will be given afterwards.

The above equation also shews that to obtain the acceleration along any line inclined at given angles to the axes, we must take the sum of the components of the accelerations parallel to the axes resolved along it. Thus the acceleration along a line inclined at angles λ, μ, ν to the axes

$$= \alpha_x \cos\lambda + \alpha_y \cos\mu + \alpha_z \cos\nu.$$

9. *When a particle moves in a plane curve to express its accelerations at any instant along and perpendicular to the radius vector.*

Let x, y be the rectangular, r, θ the polar co-ordinates, so that

$$x = r\cos\theta,$$

$$y = r\sin\theta;$$

$$\therefore \frac{dx}{dt} = \frac{dr}{dt}\cos\theta - r\sin\theta\frac{d\theta}{dt},$$

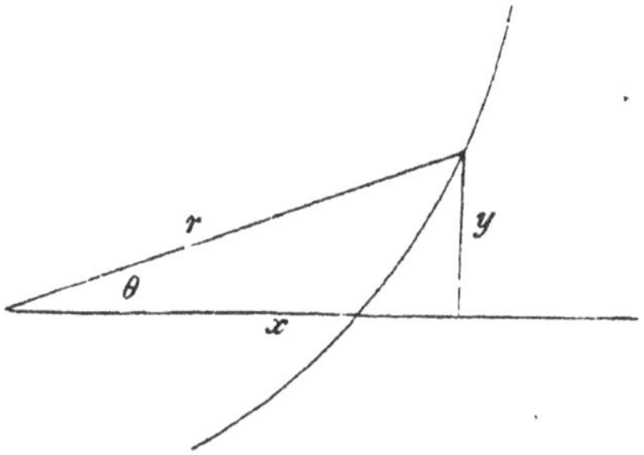

$$\frac{d^2x}{dt^2} = \left\{\frac{d^2r}{dt^2} - r\left(\frac{d\theta}{dt}\right)^2\right\}\cos\theta - \left(2\frac{dr}{dt}\frac{d\theta}{dt} + r\frac{d^2\theta}{dt^2}\right)\sin\theta.$$

Similarly,

$$\frac{d^2y}{dt^2} = \left\{\frac{d^2r}{dt^2} - r\left(\frac{d\theta}{dt}\right)^2\right\}\sin\theta + \left(2\frac{dr}{dt}\frac{d\theta}{dt} + r\frac{d^2\theta}{dt^2}\right)\cos\theta,$$

the accelerations parallel to x and y. Hence by last Art. the acceleration along the radius vector

$$= \frac{d^2y}{dt^2}\sin\theta + \frac{d^2x}{dt^2}\cos\theta$$

$$= \frac{d^2r}{dt^2} - r\left(\frac{d\theta}{dt}\right)^2.$$

And the acceleration perpendicular to the radius vector

$$=\frac{d^2y}{dt^2}\cos\theta-\frac{d^2x}{dt^2}\sin\theta$$

$$=2\frac{dr}{dt}\frac{d\theta}{dt}+r\frac{d^2\theta}{dt^2}$$

$$=\frac{1}{r}\frac{d}{dt}\left(r^2\frac{d\theta}{dt}\right).$$

10. *When a particle is in motion in any curve, to find its accelerations along, and perpendicular to, the tangent at any instant.*

Let x, y, z be the co-ordinates of the particle at the end of the time t, s the length of the arc described during that interval. Then since by the equations to the curve x, y and z are functions of s,

$$\frac{dx}{dt}=\frac{dx}{ds}\cdot\frac{ds}{dt};$$

$$\therefore\ \frac{d^2x}{dt^2}=\frac{d^2x}{ds^2}\cdot\left(\frac{ds}{dt}\right)^2+\frac{dx}{ds}\cdot\frac{d^2s}{dt^2}.$$

Similarly,

$$\frac{d^2y}{dt^2}=\frac{d^2y}{ds^2}\cdot\left(\frac{ds}{dt}\right)^2+\frac{dy}{ds}\cdot\frac{d^2s}{dt^2},$$

$$\frac{d^2z}{dt^2}=\frac{d^2z}{ds^2}\cdot\left(\frac{ds}{dt}\right)^2+\frac{dz}{ds}\cdot\frac{d^2s}{dt^2}.$$

Squaring and adding, and attending to the following equations,

$$\frac{1}{\rho^2}=\left(\frac{d^2x}{ds^2}\right)^2+\left(\frac{d^2y}{ds^2}\right)^2+\left(\frac{d^2z}{ds^2}\right)^2;$$

where ρ is the radius of curvature,

$$\left(\frac{dx}{ds}\right)^2+\left(\frac{dy}{ds}\right)^2+\left(\frac{dz}{ds}\right)^2=1,$$

$$\frac{dx}{ds}\cdot\frac{d^2x}{ds^2}+\frac{dy}{ds}\cdot\frac{d^2y}{ds^2}+\frac{dz}{ds}\cdot\frac{d^2z}{ds^2}=0,$$

we obtain

$$\left(\frac{d^2x}{dt^2}\right)^2 + \left(\frac{d^2y}{dt^2}\right)^2 + \left(\frac{d^2z}{dt^2}\right)^2 \text{ or (resultant acceleration)}^2$$

$$= \frac{1}{\rho^2} \cdot \left(\frac{ds}{dt}\right)^4 + \left(\frac{d^2s}{dt^2}\right)^2.$$

Now $\frac{d^2s}{dt^2}$ is the acceleration along the tangent, and the other part $\frac{1}{\rho} \cdot \left(\frac{ds}{dt}\right)^2$, or $\frac{v^2}{\rho}$, acts at right angles to the former as the form of the equation shews, and consequently is the acceleration perpendicular to the tangent.

From the expressions for $\frac{d^2x}{dt^2}, \frac{d^2y}{dt^2}, \frac{d^2z}{dt^2}$, we also obtain

$$\frac{d^2x}{dt^2}\left(\frac{dy}{ds} \cdot \frac{d^2z}{ds^2} - \frac{dz}{ds} \cdot \frac{d^2y}{ds^2}\right)$$

$$+ \frac{d^2y}{dt^2}\left(\frac{dz}{ds} \cdot \frac{d^2x}{ds^2} - \frac{dx}{ds} \cdot \frac{d^2z}{ds^2}\right)$$

$$+ \frac{d^2z}{dt^2}\left(\frac{dx}{ds} \cdot \frac{d^2y}{ds^2} - \frac{dy}{ds} \cdot \frac{d^2x}{ds^2}\right) = 0;$$

therefore the acceleration perpendicular to the osculating plane $= 0$. The acceleration $\frac{v^2}{\rho}$ must therefore be along a normal to the path drawn in the osculating plane.

11. It now remains to be shewn how the motion of a particle is to be determined when the accelerations to which it is subjected are given. This includes also, as will be seen, the determination of the motion when the component velocities are given.

Suppose then that α, β, γ are the given accelerations parallel to the axes, we have

$$\left.\begin{aligned}\frac{d^2x}{dt^2}&=\alpha\\ \frac{d^2y}{dt^2}&=\beta\\ \frac{d^2z}{dt^2}&=\gamma\end{aligned}\right\} \quad\ldots\ldots\ldots\ldots\ldots (1).$$

Now α, β, γ may be functions of x, y, z, t, $\frac{dx}{dt}$, $\frac{dy}{dt}$, or $\frac{dz}{dt}$, or of two or more of these quantities. Equations (1) must be solved as simultaneous differential equations if possible. Thus we have the values of $\frac{dx}{dt}$, $\frac{dy}{dt}$, $\frac{dz}{dt}$ in terms of one or more of the quantities x, y, z and t, or the component velocities are known.

Another integration if it can be performed gives x, y, and z in terms of t, and, if the latter be eliminated, we have the two equations to the path in space, and thus theoretically the motion is completely determined.

It is unnecessary to give examples of the integration of such equations, as the major part of the following chapters will be devoted to them.

12. So far for a single particle. When more particles than one are considered, Kinematics enables us to determine, from the given motions of all, their *relative* motions with respect to any one of them; or conversely, from the actual motion of one, and the motions relative to it of the others, to determine the *actual* motions of the latter in space. This depends on the following proposition.

If the velocity of any particle of a system be reversed in direction, and be communicated to each particle of the system in addition to that which it already possesses, the relative motions of all about the first thus reduced to rest, will be the same as their relative motions about it when all were in motion.

For the proof it is sufficient to notice that if at every instant the distance of two particles, and the direction of the line joining them be the same as for two other particles, the relative motions of one of each pair about the other will be the same.

Now let P and Q be the positions which the first particle and any of the others would have occupied (if undisturbed)

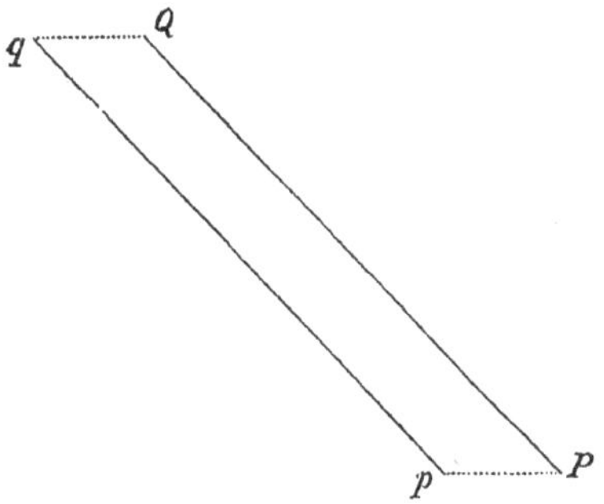

at the end of a time δt. Suppose that the same additional velocity v, in the direction Pp is impressed on both during the interval δt, then the particles will be at the end of the interval in the positions p, q, and pq is evidently equal and parallel to PQ. This will be true for all indefinitely small values of δt, and therefore at every instant the *relative* motion of the two particles p, q is the same as that of P, Q. But by impressing on the whole system a velocity equal and opposite to that of P, the latter is reduced to rest. Whence the proposition.

13. *Two particles describe similar orbits about each other and about their common center of gravity.*

Let A and B be the particles, G their center of gravity at any time. Then $\frac{AG}{GB}=\frac{B}{A}$ a constant ratio if A and B represent their masses.

The path of B about A will evidently be the same as that of A about B, since the length and direction of the

line AB are the same whichever end be supposed fixed. Also if G be fixed the path of B about it will evidently differ from that of B about A by having corresponding radii vectores diminished in the ratio $\frac{A}{A+B}$. But this is the definition of similar curves. The same of course would hold with respect to the relative path of A with respect to G. This proposition will be found of considerable use afterwards, as it enables us materially to simplify the equations of motion of two mutually attracting free particles.

14. As an instance of relative motion, suppose two particles one of which moves uniformly in a straight line while the other moves uniformly in a circle about it as center; to determine the path of the latter, the motion being in one plane.

Take the line of motion of the first as the axis of x, v its velocity, the plane of the circle as xy, a the radius of the relative circular orbit, ω the angular velocity in it. And suppose the revolving particle to be initially in the axis. Also at time t suppose the line joining the particles to be inclined at an angle θ to the axis of x. Then for the coordinates of the revolving particle we have

$$y = a \sin \theta,$$

$$x = vt + a \cos \theta.$$

But $\theta = \omega t$;

$$\therefore\ x = \frac{v}{\omega} \sin^{-1}. \frac{y}{a} + \sqrt{(a^2 - y^2)}$$

is the equation to the absolute path required. This belongs to the class of cycloids; it is prolate or curtate according as

v is greater or less than $a\omega$, or the absolute motion of the first particle greater or less than that of the other in its circular orbit. If the two are equal, or $v = a\omega$, we have the equation to the common cycloid, as is indeed evident, for the circular path may be supposed the generating circle, and the velocity of the center in its rectilinear path is equal to that of the tracing point about that center.

15. It is evident that, whatever be the relative path, if r, θ denote the relative co-ordinates of the second particle with respect to the first at time t, x, y and $\bar{x}$ the absolute co-ordinates at the same time,

$$\left.\begin{array}{l} x = \bar{x} + r\cos\theta \\ y = r\sin\theta \end{array}\right\} \text{........(1).}$$

Now in the first case, when the motion of the first body, and that in the relative orbit are given,

$\bar{x}$, r and θ are known functions of t, if therefore these values be substituted in (1), and t be eliminated, we shall have the equation between x and y, which is required.

Again, if the absolute orbits of both are given, x, y and $\bar{x}$ are known in terms of t, and thus equations (1) serve to give r and θ in terms of t, which furnishes the complete determination of the relative path, and the circumstances of its description.

16. The same may easily be extended to three dimensions.

Let $x_1\ y_1\ z_1$, $x_2\ y_2\ z_2$ be the co-ordinates in space of the particles, x, y, z the relative co-ordinates of the second with regard to the first, $u_1\ v_1\ w_1$, $u_2\ v_2\ w_2$ the velocities of each parallel to the axes, u, v, w the velocities of the second when the first is reduced to rest.

Then

$$\begin{aligned} x &= x_2 - x_1, \qquad & u &= u_2 - u_1, \\ y &= y_2 - y_1, \qquad & v &= v_2 - v_1, \\ z &= z_2 - z_1, \qquad & w &= w_2 - w_1. \end{aligned}$$

Now when the *actual* motions of the two are given all the subscribed quantities are known. Hence the above equations give the circumstances of the relative motion.

Or if the actual motion of one, and the relative motion about it of the other be known, we have $x\,y\,z$, $u\,v\,w$, $x_1y_1z_1$, $u_1v_1w_1$, to find the other six quantities for the actual motion of the other in space.

17. Some of the best illustrations of this part of our subject are to be found in what are called curves of pursuit.

These questions arose from the consideration of the path taken by a dog who in following his master always directs his course to the latter.

In order to simplify the question the rates of motion of both master and dog are supposed to continue uniform.

18. Ex. As an instance of the curve of pursuit, suppose it be required to determine the path of a point which continually with uniform velocity u moves towards another which is describing a straight line with uniform velocity v.

The curve of course is plane. Take the line of motion of the second particle Q as the axis of x, and let $\bar{x}$ denote

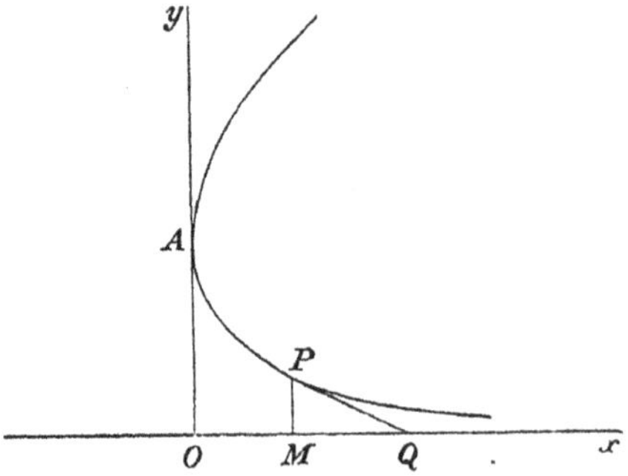

its position at the instant when the co-ordinates of the first, P, are x, y. The axis of y is chosen as that tangent to the curve of pursuit which is perpendicular to the axis of x, and the distance between the particles in that position is a.

Let $\frac{v}{u} = e$, then

$$e \, . \, AP = OQ, \text{ and } PQ \text{ is a tangent at } P.$$

These are our conditions, and lead to the following equations

$$es = \bar{x} = x - y \, . \, \frac{dx}{dy}.$$

Differentiating with respect to y, we have

$$e \, . \, \frac{ds}{dy} = - y \, . \, \frac{d^2x}{dy^2}.$$

But s increases as y diminishes,

$$\text{and } \therefore \frac{ds}{dy} = -\sqrt{\left\{1 + \left(\frac{dx}{dy}\right)^2\right\}}.$$

$$\text{Hence, } \frac{e}{y} = \frac{\frac{d^2x}{dy^2}}{\sqrt{\left\{1 + \left(\frac{dx}{dy}\right)^2\right\}}};$$

and integrating, noting that $y = a$, $\frac{dx}{dy} = 0$ together,

$$\log_e \left(\frac{y}{a}\right)^e = \log_e \left[\sqrt{\left\{1 + \left(\frac{dx}{dy}\right)^2\right\}} + \frac{dx}{dy}\right].$$

$$\text{Hence, } \left(\frac{y}{a}\right)^e = \sqrt{\left\{1 + \left(\frac{dx}{dy}\right)^2\right\}} + \frac{dx}{dy},$$

and therefore $\left(\frac{a}{y}\right)^e = \sqrt{\left\{1 + \left(\frac{dx}{dy}\right)^2\right\}} - \frac{dx}{dy}$ taking reciprocals.

And we have finally

$$2 \, . \, \frac{dx}{dy} = \left(\frac{y}{a}\right)^e - \left(\frac{a}{y}\right)^e, \quad \ldots\ldots\ldots\ldots\ldots\ldots \quad (1),$$

$$\text{or } 2(x + C) = \frac{y^{e+1}}{a^e(e+1)} + \frac{a^e}{y^{e-1}(e-1)}.$$

But $x=0$, $y=a$; $\therefore C=\frac{2ae}{e^2-1}$,

$$\text{or}\quad 2\left(x+\frac{ae}{e^2-1}\right)=\frac{y^{e+1}}{a^e(e+1)}+\frac{a^e}{y^{e-1}(e-1)} \quad \cdots\cdots \text{(2)},$$

This is true for all values of e except unity, a case to which we will presently recur.

There are two cases then of curves represented by equation (2). 1st, $e>1$, 2nd, $e<1$.

In the first case Q moves fastest, and P can never overtake it; the curve therefore never meets the axis of x, which indeed will be seen by (2) to be an asymptote.

In the second case equation (2) becomes

$$2\left(x-\frac{ae}{1-e^2}\right)=\frac{y^{1+e}}{a^e(1+e)}-\frac{a^e y^{1-e}}{1-e},$$

and for $x=\frac{ae}{1-e^2}$ we have $y=0$, and also by (1) $\frac{dx}{dy}$ infinite. Hence the curve touches the axis at this point. The rest of the curve satisfies a modified form of the question.

When $e=1$, the integral of (1) is

$$2\left(x+\frac{a}{4}\right)=\frac{y^2}{2a}-a\log_e\frac{y}{a};$$

the only case in which we do not obtain an algebraic curve. Here again the axis of x is an asymptote.

19. As an instance of relative motion let us consider the path of P with regard to Q. It will be easy to see that this corresponds exactly to the following question. A boat impelled with uniform velocity u starts from a point A in the bank of a river which runs with velocity v parallel to Qx, and tends continually to the point Q on the other bank directly opposite to A, to find its path. The constant velocity of the stream in this case communicated to P corresponds to the constant velocity of Q in the last example, but is in the opposite direction. In fact, if the earth were to be supposed moving

in the direction xQ with uniform velocity v, the river would be at rest in space, and the *actual* motions of P and Q would be the same as in the last example. So we see that in

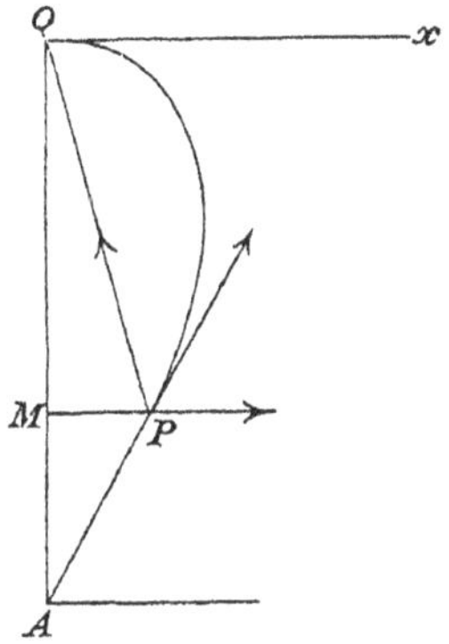

order to investigate the *relative* motions of P and Q, we have only to impress or suppose impressed on both P and Q a velocity equal and opposite to that of Q. The latter is thus reduced to rest.

To investigate the path take Q as origin, Qx, QA as the axes. Then the velocities of P are v parallel to Qx and u along PQ, and the tangent to its path is in the direction of the resultant of these two. Calling $PQx = \theta$,

$$\text{we have } \frac{dy}{dx} = -\frac{u \sin\theta}{v - u\cos\theta}$$

$$= -\frac{\sin\theta}{e - \cos\theta}$$

$$= -\frac{y}{e\sqrt{(x^2+y^2)} - x}.$$

This being a homogeneous equation is easily integrated, and we have, taking $x = 0$, $y = a$ together,

$$(1)\qquad \frac{y^{1+e}}{a^e} = \sqrt{(x^2+y^2)} - x,$$

$$\text{or } \left(\frac{\rho \sin\theta}{a}\right)^e = \frac{1-\cos\theta}{\sin\theta},$$

in polar co-ordinates. This evidently gives a parabola about Q as focus, if $e=1$.

To find the time of crossing the stream.

This may easily be effected by considering the actual velocity parallel to the axis of y,

$$\frac{dy}{dt} = -u\sin\theta$$

$$= -u \cdot \frac{y}{\sqrt{(x^2+y^2)}}.$$

Now taking quotients of y^2 by both sides of (1),

$$a^e y^{1-e} = \sqrt{(x^2+y^2)} + x;$$

$$\therefore\ 2\sqrt{(x^2+y^2)} = a^e y^{1-e} + a^{-e} y^{1+e};$$

$$\therefore\ \frac{dy}{y}\left(a^e y^{1-e} + a^{-e} y^{1+e}\right) = -2u\,dt,$$

and taking the integral from a to 0, calling T_1 the time of crossing,

$$\frac{a}{1-e^2} = u \cdot T_1;\quad \therefore\ T_1 = \frac{au}{u^2-v^2}.$$

But if there had been no current we should have had for the time of crossing,

$$T_0 = \frac{a}{u};\quad \therefore\ \frac{T_1}{T_0} = \frac{u^2}{u^2-v^2}.$$

In the integration of course $e<1$ else the boat would never reach Q.

If $e=1$, the boat will reach the bank but not at Q, and if $e>1$ it never reaches the bank at all.

20. DEF. *If the motion of a particle in a plane be considered with reference to a fixed point in that plane, the angular velocity of the former about the latter means the rate of increase of the angle made by the line joining the two, with some fixed line in the plane.*

Suppose this angle to be represented by θ at time t; then at time $t+\delta t$ it has the value $\theta+\delta\theta$, and it may be shewn as before (Art. 2), that if ω represent the angular velocity required, then

$$\omega=\frac{d\theta}{dt}.$$

Ex. As an illustration, suppose a particle move uniformly with velocity v in a straight line; to find at any instant its angular velocity about a point whose distance from the straight line is a.

Taking as initial line the perpendicular from the point on the line of motion; the polar equation of the latter is

$$r=a\sec\theta.$$

Also if when $t=0$, $\theta=0$, we have

$$r\sin\theta=vt.$$

Hence, $a\tan\theta=vt$,

$$\text{and } \therefore\ \omega=\frac{d\theta}{dt}=\frac{va}{a^2+v^2t^2}=\frac{va}{r^2}.$$

We will take another example of the estimation of angular velocity.

21. A point describes a circle with uniform velocity; it is required to find the actual velocity, and the angular velocity (about the center) in any orthographic projection.

Let ApA' be an ellipse and APA' the auxiliary circle. Then the former will correspond to the projection if its axes

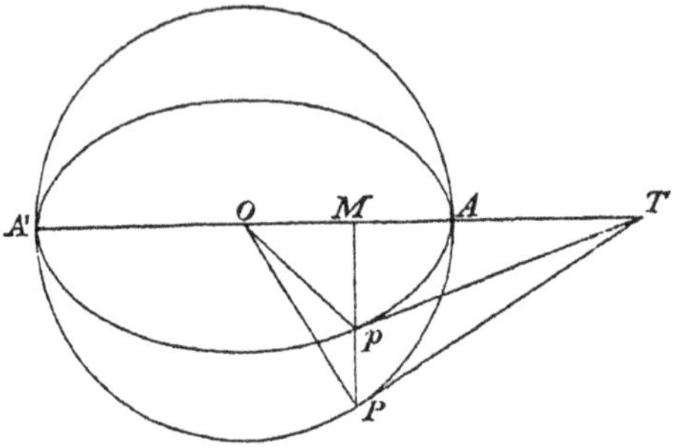

be made in the ratio of the cosine of the angle (α) between the planes. Also if PpM be perpendicular to AA', P and p will be corresponding points in the two. Draw the tangents pT, PT; then

$$\frac{\text{actual velocity at } p}{\dots\dots\dots\dots\dots\dots \ P} = \frac{pT}{PT}, \quad \text{and if } TOP = \theta,$$

$$\frac{\text{velocity at } p}{\dots\dots\dots \ P} = \frac{\sqrt{(PT^2 \sin^2\theta + PT^2 \cos^2\theta \cos^2\alpha)}}{PT}$$

$$= \sqrt{(\sin^2\theta + \cos^2\theta \cos^2\alpha)}$$

$$= \sqrt{(1 - \sin^2\alpha \cos^2\theta)}.$$

Now if $TOp = \phi$, the $\dfrac{\text{angular velocity at } p}{\dots\dots\dots\dots\dots\dots \ P} = \dfrac{d\phi}{d\theta}$

$$= \frac{d}{d\theta} \tan^{-1} (\cos\alpha \tan\theta)$$

$$= \frac{\cos\alpha}{\cos^2\theta + \cos^2\alpha \sin^2\theta}$$

$$= \frac{\cos\alpha}{1 - \sin^2\alpha \sin^2\theta}.$$

This is a maximum when $\theta = \frac{\pi}{2}$ when its value is $\sec \alpha$,

.... minimum .... $= 0$ .... $\cos \alpha$;

$\therefore$ if ω_1 and ω_2 be the greatest and least angular velocities in the projection,

$\sqrt{(\omega_1 \omega_2)}$ is the angular velocity in the original path.

22. The principles already enunciated, and the examples given of their application, will suffice for the solution of problems on this part of the subject.

Other examples of the application of these principles, such as the kinematical part of the investigations of the Hodograph, Tractory, &c., will be more appropriately introduced in future chapters.

EXAMPLES.

(1) A particle moves from rest in a given path, and its velocity at any instant is proportional to the time elapsed since its motion commenced; find the space described in a given time.

(2) If a particle begin to move with velocity v, and at equal intervals of time a velocity u be communicated to it in the same direction; find the space described in n such intervals.

(3) If the position of a particle moving in a plane be determined by the co-ordinates ρ and ϕ, ρ being measured from a fixed circle (radius a) along a tangent which has revolved through an angle ϕ from a fixed tangent; investigate the following expressions for the accelerations along and perpendicular to ρ respectively,

$$\frac{d^2\rho}{dt^2} - \rho\left(\frac{d\phi}{dt}\right)^2 + a\frac{d^2\phi}{dt^2}$$

$$\text{and } \frac{1}{\rho}\frac{d}{dt}\left(\rho^2\frac{d\phi}{dt}\right) + a\left(\frac{d\phi}{dt}\right)^2.$$

(4) The velocity of a particle parallel to each axis is proportional to the product of the other two co-ordinates; what are the equations to the path, and what is the time of describing a given portion when the curve passes through the origin?

(5) A particle moves in a plane, its velocities parallel to the axes of x and y are

$$u + ey \text{ and } v + \epsilon x \text{ respectively,}$$

shew that it moves in a conic section.

(6) Two particles are moving with uniform velocity in two straight lines, 1st in a plane, 2nd in space; given the initial circumstances, find when they are nearest to each other. Shew also that in both cases the relative path is a straight line, described with uniform velocity.

(7) A number of particles are moving with uniform velocity in straight lines in space; determine the motion of their common center of gravity.

(8) A cannon-ball is moving in a direction making an acute angle θ with a line drawn from the ball to an observer; if V be the velocity of sound, and nV that of the ball, prove that the whizzing of the ball at different points of its course will be heard in the order in which it is produced, or in the reverse order, according as $n <> \sec\theta$.

(9) Shew that any indefinitely small motion given to a plane figure in its own plane is equivalent to a rotation through an indefinitely small angle about some point in the figure.

(10) The highest point of the wheel of a carriage rolling on a horizontal road moves twice as fast as each of two points in the rim whose distance from the ground is half the radius of the wheel.

(11) A rod of given length moves with its extremities in two given lines which intersect; shew how to draw a tangent to the path described by any point of the rod.

(12) Investigate the position of the instantaneous center about which the rod is turning, and apply this also to solve the preceding question.

(13) One circle rolls on another whose center is fixed. From the initial and final positions of a diameter in each determine how much of their circumferences have been in contact.

(14) In the Example of Art. (18) find in the case of $e=1$ the ultimate distance of the particles, and for $e<1$ the length of time occupied in the pursuit.

(15) In the example of Art. (19) find the greatest distance which the boat is carried down the stream, and shew that when it is at that point its velocity is $\sqrt{(u^2-v^2)}$.

When $u=v$, shew directly that the curve described is a parabola.

(16) The velocity of a stream varies as the distance from the nearest bank; shew that a man attempting to swim directly across will describe two semiparabolas. Find by how much the mean velocity is increased.

(17) A particle moves uniformly in a circle; find an expression for its angular velocity about any point in the plane of the circle.

(18) If the velocity of a particle moving in a plane curve vary as the radius of curvature, shew that the direction of motion revolves with uniform angular velocity.

(19) Two bevilled wheels roll together; having given the angular velocity of the first wheel and the inclinations of the axes of the cones, find their vertical angles that the second may revolve with given angular velocity.

(20) Supposing the Earth and Venus to describe in the same plane circles about the Sun as center; investigate an

expression for the angular velocity of the Earth about Venus in any position, the actual velocities being inversely as the square roots of their distances from the Sun.

(21) A particle moving uniformly round the base of an oblique cone is projected by generating lines on a subcontrary section; find its angular velocity about the center of the latter.

CHAPTER II.

LAWS OF MOTION.

23. HAVING, in the preceding chapter, considered the purely geometrical properties of the motion of a particle, we must now treat of the causes which produce various circumstances of motion, and the experimental laws, on the assumed truth of which all our succeeding investigations are founded.

When the velocity, or direction of motion of a body, or both, change, or when a body at rest begins to move, we naturally attribute this to some extraneous cause. This cause, of whatever nature it may be, is known by the general name of *Force*. Hence the definition,

Force is any cause which produces, or tends to produce, a change in a body's state of rest or motion.

24. A particle of matter is supposed to be so small, that though retaining its material properties, it may be treated as far as its co-ordinates, &c. are concerned, as a geometrical point.

It is an experimental fact, that a particle of matter possesses in itself no power of setting itself in motion if at rest, or of changing the direction or velocity of its motion if it be moving. This property is named *Inertia:* and the preceding result of experiment is generally presented in the following form; which is called the *First Law of Motion.*

A particle acted on by no external forces will either remain at rest or move uniformly in a straight line.

This may be considered self-evident, but it was long in being recognized as a principle, and the want of it was almost a complete barrier to the progress of the science.

We may notice that the more nearly we satisfy the condition of a particle's freedom from the action of external forces the more nearly is its motion found to be rectilinear and uniform; but the most satisfactory experimental proof of this and the other law of motion is to be found in the complete agreement of calculations founded on them, with observed motions, especially in the case of the bodies composing the Solar System.

25. Hence if the velocity of a particle be observed to vary (supposing for simplicity's sake that its direction of motion is unchanged) we are to conclude that it is under the action of some force. How is this to be measured? Before entering upon this inquiry it will be advisable to consider the effect of a force or forces on a *moving particle*, since as every atom in the universe is in motion, it is only under such circumstances that we can study practically the effect of force.

To find the effect on a particle in motion, of a force either singly or in combination with other forces, we must again appeal to experiment. And the result to which this leads us, is the *Second Law of Motion.*

If any number of forces act upon a particle in motion, the effect of each force, in respect to intensity and direction, in altering the particle's motion, is the same as if it had acted singly on the particle at rest.

Various phenomena, continually occurring around us, furnish more or less satisfactory arguments for the truth of this law. The fall of a body from the mast head of a vessel to the same point of the deck however rapidly the vessel be moving; the equal ease of walking along or across a saloon carriage running on a smooth railway; and various other instances, are commonly cited. An excellent illustration is furnished by the fact of a pendulum performing its vibrations in the same time whether its plane of motion be that of the meridian of a place, or one inclined to it.

26. Def. *Momentum is the product of the mass and the velocity of a body.*

The mass of a body is the quantity of matter which it contains. Suppose for instance a gas be compressed into half the space it would occupy at the ordinary atmospheric pressure, the mass of a cubic inch of it in the former state would be double of that of a cubic inch in the latter; and the density would be said to be doubled. By such reasoning we come to the conclusion that if V be the volume and ρ the density of an homogeneous body the expression for its mass is $V\rho$; if the body be heterogeneous this will become $\iiint \rho dV$. But in order to find ρ, we are driven to some such expedient as seeking the specific gravity. The weight of a body therefore is the ultimate standard of reference for the mass. We shall however for the present assume that its value is known, and proceed with the consideration of the effect of a given force on a given mass.

27. *The measure of a force if uniform is the momentum generated by it in an unit of time; if variable, that which would have been generated by the force if at any instant it had continued uniform for an unit of time.*

Now by the second law of motion it is evident that two equal forces acting in the same direction will produce in an unit of time double the velocity which would have been produced by either singly. And so for other equimultiples of the force and the velocity.

Hence F being the force and v the velocity it produces in an unit of time in a mass M,

$$F \propto Mv \text{ when } M \text{ is constant.}$$

But if we were to divide the mass M and the force F into the same number of equal parts, and suppose each of the parts of M to be acted on by one of the parts of F, the velocity produced in each would evidently be equal to v, and therefore

$$F \propto Mv, \text{ when } v \text{ is constant.}$$

Hence, $F = CMv$, where C is some constant. The force therefore is *proportional to* the momentum it generates in an unit of time, and if we call the unit of force that which generates an unit of momentum in an unit of time, we have $C = 1$, or

$$F = Mv, \quad \ldots\ldots\ldots\ldots\ldots (1).$$

Hence the momentum generated in time t by an uniform force F is Ft.

If F be variable, it is easy to see that by the principles of the first chapter (Art. 7) equation (1) becomes

$$F = M\frac{dv}{dt} = M\alpha \quad \ldots\ldots\ldots\ldots (2),$$

where α is the acceleration which would have been produced in an unit of time had the force remained constant for that period. Also the momentum generated in time t by such a force F will be $\int_0^t F dt$.

28. It is found by experiments made on the fall of bodies in the exhausted receiver of an air-pump, that of whatever substance these bodies may be, they fall through 16.1 feet in one second, and have at the end of that time a velocity of 32.2 feet per second. This is denoted by the letter g. It is also found that the velocity of a falling body increases proportionally to the time, and therefore by the result of equation (1), the Earth's attraction is an uniform force (at least within small distances from the surface.) If a body be prevented from falling by the tension of a string, the value of that tension is the weight of the body (or the force which would urge it in its fall were the string cut); and if this latter equal W, and if a second be taken as the unit of time, (1) becomes

$$W = Mg.$$

If we take the ounce as the unit of weight, we know that the weight of a cubic foot of distilled water (whose specific gravity is unity) is 1000 oz. Therefore since unit of volume of a substance whose density is unity is unit of mass,

$$1000 = 32.2 \times \text{weight of unit of mass},$$

Hence the unit of mass of any substance is such that its weight is

$$\frac{1000}{32.2} \text{ oz.}$$

29. It is worthy of notice, that if F be measured by its effects on an unit of mass, (2) gives

$$F = a, \quad \ldots\ldots\ldots\ldots\ldots (3),$$

or the value of the force is simply the acceleration. Measured in this way it is called accelerating force, in the way we have here adopted it is called moving force. It is important to observe, that the difference lies entirely in the manner in which we measure the cause by the effect, and not in any diversity between various kinds of force.

30. Difficulty is sometimes felt by the student in conceiving how such a force as gravity produces a like effect on all bodies however different their masses; but such will easily be got over by considering a body divided into any number of parts, and falling to the ground. All these parts will fall together, and will have at any instant the velocity which the mass itself would have had, had it not been broken up. Such a force as gravity is exerted with the same intensity on all equal particles of matter at nearly the same distance from the Earth, and when a number of these are joined together the really acting force will be proportional to the number of particles. This is very different from most artificial or animal forces, such as pressures, tensions, forces produced by explosions, &c. which have a definite value, and whatever body they act on can produce but a fixed amount of momentum.

31. Accordingly therefore as we consider or not the mass of a body, equation (2) or (3) combined with the second law of motion and with the principles explained in the first chapter for the resolution of accelerating effects, will give us the equations of motion of a body under the action of given forces.

Suppose R, S, T to be the sums of the resolved parts, parallel to the axes, of the forces acting on a particle of mass

M; we know that by the second law of motion we may consider the effects of these separately, hence we have by (2) the three equations of motion

$$\left.\begin{aligned} M\frac{d^2x}{dt^2} &= R, \\ M\frac{d^2y}{dt^2} &= S, \\ M\frac{d^2z}{dt^2} &= T. \end{aligned}\right\}$$

And if we wish the equations of motion under the action of the same forces along a given curve of which s is the arc, we have by (2) and by Art. (8) the equation

$$M.\frac{d^2s}{dt^2} = R.\frac{dx}{ds} + S.\frac{dy}{ds} + T.\frac{dz}{ds}.$$

32. This is true as long as the forces are finite. To consider the case when the force is indefinitely great, let us take for simplicity the acceleration in one direction only, as that of x.

Here $M\dfrac{d^2x}{dt^2} = R$, and for the momentum generated in an indefinitely small time τ we have the expression

$$M\frac{dx}{dt} = Mv = \int_0^\tau Rdt.$$

Now τ being indefinitely small and R indefinitely great, it *may* happen that the value of the integral is finite. This case we will consider under the head *Impact*. It may be as well to notice at present that such cases may be treated entirely by themselves, since no finite force can in an indefinitely short time generate a finite velocity. In fact if R_1 be some value intermediate to the greatest and least that R a finite force has in the interval from 0 to τ, we have

$$\int_0^\tau Rdt = R_1\tau.$$

But one of the factors is indefinitely small, and the other is by hypothesis finite; hence the above.

CHAPTER III.

RECTILINEAR MOTION.

33. THE simplest case of motion which we have to consider is that of a particle in a straight line. This may be due to the accelerating force acting at every instant in the direction of motion; or the particle may be supposed to be constrained to move in a straight line by its being enclosed in a straight tube of indefinitely small bore.

34. Let us consider first the case in which the resultant accelerating force is in the line of motion.

Let m be the mass of the particle, P its position at any time t, f the resultant acceleration acting always along OP, O being a fixed point in the line of motion.

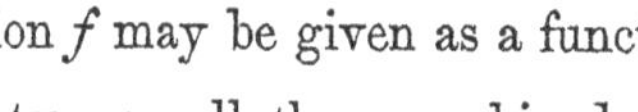

Let $OP = x$, then the equation of motion is

$$m \cdot \frac{d^2x}{dt^2} = mf.$$

In this equation f may be given as a function of x, of $\frac{dx}{dt}$, or of t, or of any two or all three combined, but in any case the first and second integrals of the equation (if they can be obtained) will give x and $\frac{dx}{dt}$ in terms of t; that is, the position and velocity of the particle at any instant will be known.

The only one of these cases we will now consider is that in which f is given as a function of x; those in which f is a function of $\frac{dx}{dt}$ or of $\frac{dx}{dt}$ and x being reserved for the Chapter on Motion in a Resisting Medium; and those in which f involves t explicitly, possessing no interest as they never occur in nature.

The simplest supposition then is that f is constant.

35. PROP. *A particle projected from a given point with a given velocity is acted on by a constant force in the line of its motion; to determine the position and velocity of the particle at any time.*

Let A be the initial position of the particle, P its position at any time t, v its velocity at that time, and f the constant

accelerating force acting on it. Take any fixed point O in the line of motion as origin, and let $OA = a$, $OP = x$. The equation of motion becomes

$$\frac{d^2x}{dt^2} = f \text{ .... } (1).$$

Integrating

$$\frac{dx}{dt} = v = ft + C,$$

C being a constant to be determined by the initial circumstances of motion. Suppose the particle projected from A in the positive direction with the velocity V, then when $t = 0$, $v = V$; $\therefore\ C = V$, and

$$\therefore\ \frac{dx}{dt} = v = V + ft \text{ .... } (2).$$

Integrating again

$$x = C' + Vt + f\frac{t^2}{2}.$$

But when $t = 0$, $x = a$; $\therefore\ C' = a$, and

$$\therefore\ x = a + Vt + f\frac{t^2}{2} \text{ .... } (3).$$

Equations (2) and (3) give the velocity and position of the particle in terms of t, and the velocity may be determined in terms of x by eliminating t between them, but the same result will be obtained more directly by multiplying (1) by $2\frac{dx}{dt}$ and integrating. This gives

$$\left(\frac{dx}{dt}\right)^2 = v^2 = C'' + 2fx.$$

But when $x = a$, $v = V$; $\therefore$ $C'' = V^2 - 2fa$, and

$$\therefore\ v^2 = V^2 + 2f(x - a) \text{ .... } (4).$$

36. The most important case of the motion of a particle under the action of a constant force in its line of motion is that in which the force is gravity. For the weight of bodies in the same latitude at small distances above the Earth's surface may be considered constant, and therefore if we denote the accelerating force of gravity by g, and consider the particle projected vertically downwards, equations (2), (3), (4) of last Article become

$$\left.\begin{aligned} v &= V + gt \\ x &= a + Vt + \tfrac{1}{2}gt^2 \\ v^2 &= V^2 + 2g(x - a) \end{aligned}\right\} \text{ .... } (A),$$

x being measured as before from a fixed point O in the line of motion. As a particular instance consider the particle dropped from O, then A coincides with O and $a = 0$, $V = 0$;

$$\therefore\ v = gt \text{ .... } (1),$$

$$x = \tfrac{1}{2}gt^2 \text{ .... } (2),$$

$$v^2 = 2gx \text{ .... } (3).$$

37. As another particular instance consider the particle projected vertically upwards. Here it must be remembered that if we measure x upwards from the point of projection, the force tends to diminish x and the equation of motion is

$$\frac{d^2x}{dt^2} = -g.$$

In other respects the solution is the same. Taking $a=0$ then in equation (A) and changing the sign of g, we obtain

$$v = V - gt \qquad (4),$$

$$x = Vt - \frac{gt^2}{2} \qquad (5),$$

$$v^2 = V^2 - 2gx \qquad (6).$$

Equation (4) shews that the velocity continually diminishes and becomes zero when $t = \frac{V}{g}$, and (6) that the height corresponding to $v = 0$, or the height to which the particle will ascend $= \frac{V^2}{2g}$. After this the velocity becomes negative or the particle begins to descend, and (5) shews that it will return to the point of projection when $t = \frac{2V}{g}$ as then $x = 0$, and the velocity with which it returns to that point will by (6) be equal to the velocity of projection.

38. Prop. *A particle descends a smooth inclined plane under the action of gravity, the motion taking place in a vertical plane perpendicular to the intersection of the inclined with the horizontal plane; to determine the motion.*

Let M be the mass of the particle, P its position at any time t on the inclined plane OA, OP its distance from a fixed

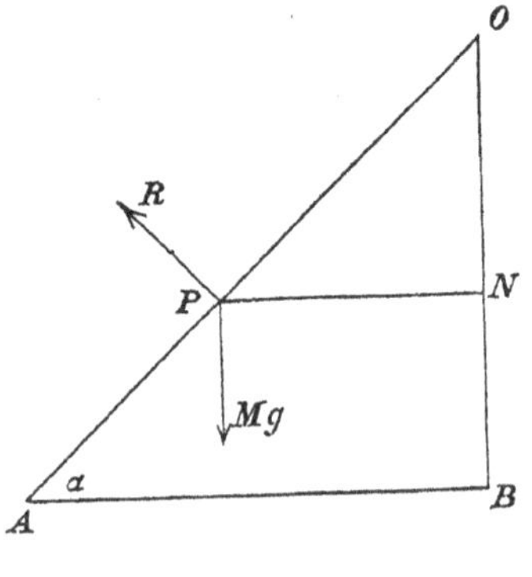

point O in the line of motion $= x$, and let α be the inclination of OA to the horizontal line AB. The only impressed force on the particle is its weight Mg vertical, and this may be resolved into two, $Mg \sin \alpha$ along, and $Mg \cos \alpha$ perpendicular to OA. Besides these there is the unknown force R or reaction of the plane perpendicular to OA arising from the pressure of the particle, but neither this nor the component $Mg \cos \alpha$ can affect the motion along the plane. The equation of motion is therefore

$$\frac{d^2x}{dt^2} = g \sin \alpha,$$

the solution of which as $g \sin \alpha$ is constant, is included in that of the proposition in Art. 35, and all the results for particles moving vertically under the action of gravity will be made to apply to it by writing $g \sin \alpha$ for g. Thus if the particle started from rest at O, we get from equations (1), (2), (3) of Art. 36 by this means,

$$v = g \sin \alpha \,.\, t \quad\quad (1),$$

$$x = \tfrac{1}{2} g \sin \alpha \,.\, t^2 \quad\quad (2),$$

$$v^2 = 2g \sin \alpha \,.\, x \quad\quad (3).$$

39. Equation (3) proves an important property with regard to the velocity acquired at any point of the descent. For draw PN parallel to AB, meeting the vertical line through O in N, then if v be the velocity at P, we have

$$v^2 = 2g \,.\, \sin \alpha \,.\, OP$$

$$= 2g \,.\, ON.$$

Comparing this with equation (3) of Art. 36, we see that the velocity at P is the same as that which a particle would acquire by falling freely from rest through the vertical distance ON; in other words the velocity at any point, of a particle sliding down a smooth inclined plane, is that due to the vertical height through which it has descended.

40. From (2) again we derive immediately the following

PROP. *The time of descent down all chords drawn through the highest or lowest point of a vertical circle is the same.*

Let AB be the vertical diameter of the circle, AC any

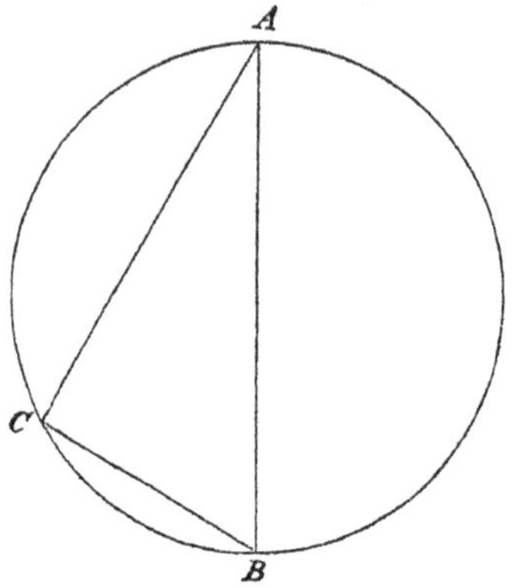

chord through A; join BC; then if T be the time of descent down AC, we have by equation (2),

$$AC = \tfrac{1}{2} g T^2 \cos BAC.$$

But $AC = AB \cos BAC$;

$$\therefore\ AB = \tfrac{1}{2} g T^2,$$

$$\text{or } T = \sqrt{\left(\frac{2AB}{g}\right)},$$

which being independent of the position of the chord, gives the same time of descent for all.

It may similarly be shewn that the time of descent down all chords through B is the same.

41. In Art. 38 we have considered the inclined plane smooth, but the motion will still be uniformly accelerated when the plane is rough. For since there is no motion perpendicular to OA (see fig. Art. 38), we must have

$$R = Mg \cos \alpha.$$

If μ then be the coefficient of dynamical friction, which as we shall afterwards see is independent of the velocity of the particle, the retarding force of friction will be μR or $\mu Mg \cos \alpha$, and the equation of motion will become

$$\frac{d^2x}{dt^2} = g \sin \alpha - \mu g \cos \alpha,$$

the second member still being constant, and the solution therefore similar to those we have already considered.

42. PROP. *A particle moves under the action of a force in the line of motion varying directly as the distance of the particle from a fixed point in that line, to determine the motion.*

Let O be the fixed point, P the position of the particle at any time t, v its velocity at that time, and let $OP = x$. Then

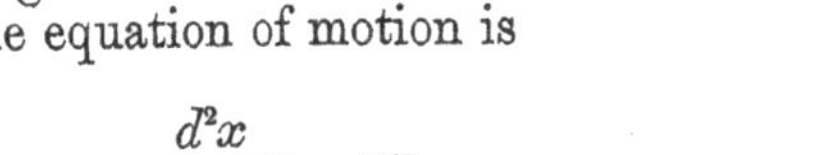

if μ be the accelerating force on a particle at a unit of distance from O, and which is supposed known, the force at P will be μx, and being directed towards O will tend to diminish x. Therefore the equation of motion is

$$\frac{d^2x}{dt^2} = -\mu x,$$

$$\text{or } \frac{d^2x}{dt^2} + \mu x = 0 \quad \text{...........} \quad (1).$$

Multiplying this equation by $2\frac{dx}{dt}$, and integrating, we obtain

$$\left(\frac{dx}{dt}\right)^2 = \mu (A^2 - x^2) \quad \text{...........} \quad (2),$$

$$\text{or } \frac{dt}{dx} = \frac{1}{\sqrt{(\mu)}} \cdot \frac{1}{\sqrt{(A^2 - x^2)}},$$

A being the constant introduced in the integration. Integrating again

$$\sqrt{(\mu)} \,.\, t + B = \cos^{-1}\frac{x}{A};$$

$$\text{or } x = A \cos \{\sqrt{(\mu)} \,.\, t + B\} \;\ldots\ldots\; (3),$$

the complete integral of (1) involving two arbitrary constants A and B, the values of which are to be determined from the initial distance, and velocity of projection. Thus from (3),

$$\frac{dx}{dt} = v = -\sqrt{(\mu)} \,.\, A \sin \{\sqrt{(\mu)} \,.\, t + B\} \;\ldots\; (4).$$

43. Suppose the particle projected from A in the positive direction with the velocity V, and let $OA = a$: then when $t = 0$, $x = a$, $v = V$; and therefore from (3) and (4)

$$a = A \cos B,$$

$$V = -\sqrt{(\mu)} \,.\, A \sin B,$$

which determine A and B, and then (3) and (4) give the position and velocity of the particle at any instant. The velocity in terms of x is obtained most directly from (2), for when $x = a$, $v = V$; $\therefore\ V^2 = \mu\,(A^2 - a^2)$, and

$$\therefore\ v^2 = V^2 + \mu\,(a^2 - x^2).$$

44. Equations (3) and (4) give periodical values of x and v, so that all the circumstances of motion will be the same at the time $t + \dfrac{2\pi}{\sqrt{(\mu)}}$ as at the time t. They also shew that the velocity becomes zero when $\sqrt{(\mu)} \,.\, t + B = 0$, and that the corresponding value of x is the greatest possible. Hence the particle will move in the positive direction to a distance A from O, then begin to return, and since when $\sqrt{(\mu)}\, t + B = \pi$, $v = 0$ again, and $x = -A$, it will pass through O, move to an equal distance on the other side, and so on, the time of a complete oscillation, that is, the time from its leaving any point until it passes through it again in the same direction,

being $\frac{2\pi}{\sqrt{(\mu)}}$. This is a remarkable result, as it shews that the time of oscillation is independent of the velocity and distance of projection, and depends solely on the intensity of the force.

The above proposition includes the motion of a particle within an homogeneous sphere, in a straight line to the center. For the attraction of such a sphere on a particle within it is proportional to its distance from the center, and the equation of motion will consequently be the same as that we have been considering.

45. If the force in Art. (42) be supposed repulsive or directed always from the center instead of towards it, the equation of motion becomes

$$\frac{d^2x}{dt^2} = \mu x,$$

the integral of which is known to be

$$x = A\epsilon^{\sqrt{(\mu)}t} + B\epsilon^{-\sqrt{(\mu)}t};$$

and therefore x constantly increases with t, or the motion is not oscillatory.

46. It is to be remarked that we cannot always apply the same equation of motion to the negative and positive sides of the origin as we have done in the case of Art. 42. The fact of our being able to do so arises from the force itself μx changing signs with x, for by looking at the figure it will be seen that when x is negative the force tends to increase x algebraically, and the equation ought properly to be written

$$\frac{d^2x}{dt^2} = +\mu(-x).$$

In general when the law of force is proportional to the nth power of the distance the equations of motion for the positive and negative sides of the origin are respectively

$$\frac{d^2x}{dt^2} = \mu x^n,$$

$$\text{and } \frac{d^2x}{dt^2} = -\mu(-x)^n.$$

The only case therefore when the same equation of motion will apply to both sides of the origin is when n is of the form $\frac{2m+1}{2m'+1}$, where m, m' are any whole numbers including zero, as then

$$-(-x)^n = x^n.$$

47. In all other cases the investigation of the motion will generally consist of two parts, one for each side of the origin; and in one case even when n is of the form $\frac{2m+1}{2m'+1}$ it is necessary to consider these parts separately on account of the form of the integral not being sufficiently general to include both. This is when $m=0$ and $m'=-1$, as then the equation of motion becomes

$$\frac{d^2x}{dt^2} = -\frac{\mu}{x}.$$

Multiplying this by $2\frac{dx}{dt}$ and integrating we have

$$\left(\frac{dx}{dt}\right)^2 = C - 2\mu \log x,$$

which becomes impossible when x is negative. But it is evident that we may then write the integral

$$\left(\frac{dx}{dt}\right)^2 = C - 2\mu \log(-x),$$

which is therefore the proper form for the negative side of the origin. These equations cannot generally be integrated farther, but we will shew towards the end of the Chapter how the time of reaching the origin may be determined.

48. PROP. *A particle constrained to move in a straight line, is acted on by a force always directed to a point outside the line,*

and varying directly as the distance of the particle from that point, to determine the motion.

The constraint here contemplated may be conceived by considering the particle either as an indefinitely small ring sliding on a smooth rod, or as a material particle sliding in a straight tube of indefinitely small bore.

Let AB be the straight line, P the position of the particle

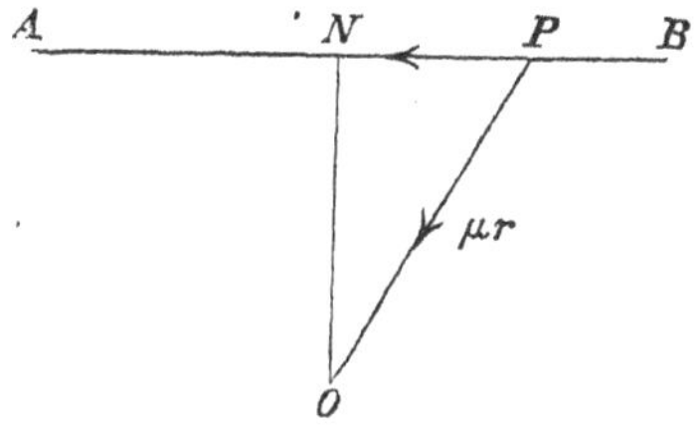

at any time, O the point to which the force on P is always directed. Draw ON perpendicular to AB, and let $NP = x$; then if $OP = r$, and μ as formerly be the force at an unit of distance, the force on P along $PO = \mu r$. This may be resolved into two, one along and the other perpendicular to AB, of which the latter has no effect on the motion of the particle, which is therefore due only to the accelerating force

$$\mu r \cos OPN = \mu x, \text{ along } PN.$$

The equation of motion is therefore

$$\frac{d^2x}{dt^2} = -\mu x,$$

the same as in Art. (42). The motion of the particle will therefore be oscillatory about N, the time of a complete oscillation being $\frac{2\pi}{\sqrt{(\mu)}}$, and all the circumstances of motion the same as for a free particle moving in AB under the action of the same force always directed towards N.

49. PROP. *A particle moves in a straight line under the action of a force always directed to a point in that line and*

varying inversely as the square of the distance from that point; to determine the motion.

Let as before O be the fixed point, P the position of the

particle at the time t, $OP=x$; then the equation of motion is

$$\frac{d^2x}{dt^2}=-\frac{\mu}{x^2}.$$

Multiplying by $2\frac{dx}{dt}$ and integrating, we get

$$\left(\frac{dx}{dt}\right)^2=v^2=C+\frac{2\mu}{x}.$$

Suppose the particle to start from rest at a point A distant a from O, then when $x=a$, $v=0$;

$$\therefore\ C=-\frac{2\mu}{a},\text{ and}$$

$$\therefore\ \left(\frac{dx}{dt}\right)^2=v^2=2\mu\left(\frac{1}{x}-\frac{1}{a}\right)\ \ldots\ldots\ldots\ldots (1),$$

which gives the velocity of the particle at any distance x from the origin. Again from (1)

$$\frac{dx}{dt}=-\sqrt{(2\mu)}\sqrt{\left(\frac{a-x}{ax}\right)},$$

the negative sign being taken, since in the motion towards O, x diminishes as t increases;

$$\therefore\ \frac{dt}{dx}=-\sqrt{\left(\frac{a}{2\mu}\right)}\cdot\frac{x}{\sqrt{(ax-x^2)}}$$

$$=\sqrt{\left(\frac{a}{2\mu}\right)}\left\{\frac{1}{2}\cdot\frac{a-2x}{\sqrt{(ax-x^2)}}-\frac{a}{2}\cdot\frac{1}{\sqrt{(ax-x^2)}}\right\}.$$

Integrating

$$t=\sqrt{\frac{a}{2\mu}}\left\{\sqrt{(ax-x^2)}-\frac{a}{2}\text{vers}^{-1}\frac{2x}{a}+C'\right\}.$$

Now when $t=0$, $x=a$, and therefore $C'=\frac{a\pi}{2}$;

$$\therefore \sqrt{\frac{2\mu}{a}}\,.\,t=\sqrt{(ax-x^2)}-\frac{a}{2}\text{vers}^{-1}\frac{2x}{a}+\frac{a\pi}{2},$$

which is the relation between x and t.

50. By putting $x=0$ we find that the time of arriving at $O=\frac{\pi}{2}\sqrt{\left(\frac{a^3}{2\mu}\right)}$, and (1) shews that the velocity there is infinite. On this account we are precluded from applying our formulæ to determine the motion after arriving at O, but it is to be observed that although at any point very near to O there is a very great force tending towards O, at the point O itself there is no force at all, and therefore the particle approaching the center of force with an indefinitely great velocity must pass through it. Also everything being the same at equal distances from the center, we may consider the motion checked as rapidly as it was generated, and therefore the particle will proceed to a distance on the other side of S equal to that from which it started. The motion will then continue oscillatory.

51. The above case of motion includes that of a body falling from a great height above the Earth's surface. For a sphere attracts an external particle with a force varying inversely as the square of the distance of the particle from the center, and therefore if x be the distance of a body from the Earth's center, R the Earth's radius, and g the accelerating force of gravity at the Earth's surface, the equation of motion will be

$$\frac{d^2x}{dt^2}=-g\frac{R^2}{x^2},$$

the same equation as before, writing μ for gR^2. The results therefore just obtained will apply to this case. Thus if we wish to find the velocity which a body would acquire in falling from a height a to the Earth's surface, we have from (1), putting $\mu = gR^2$,

$$v^2 = 2gR^2\left(\frac{1}{x} - \frac{1}{a}\right);$$

and therefore if V be the velocity when $x = R$ or the required velocity,

$$V^2 = 2gR\,\frac{a-R}{a}.$$

52. *A particle is constrained to move in a straight line, and is acted on by a force always directed to a point outside that line and varying inversely as the square of the distance; to determine the motion.*

Let AB be the straight line, P the position of the particle at any time, O the point to which the force is always directed,

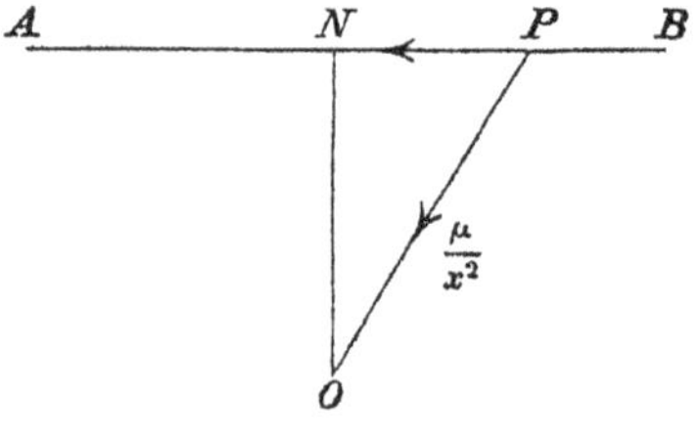

μ the force at an unit of distance. Draw ON perpendicular to AB and let $ON = b$, $NP = x$; then the accelerating force on P along $PO = \frac{\mu}{PO^2}$, and as in Art. (48), the only part of this which produces motion is the resolved part along PN. Therefore the equation of motion is

$$\frac{d^2x}{dt^2} = -\frac{\mu}{OP^2}\cos OPN$$

$$= -\frac{\mu x}{(x^2+b^2)^{\frac{3}{2}}} \text{ .... } (1).$$

Multiplying by $2\frac{dx}{dt}$ and integrating

$$\left(\frac{dx}{dt}\right)^2 = v^2 = C + \frac{\mu}{(x^2+b^2)^{\frac{1}{2}}},$$

where C is to be determined in the usual manner.

53. This equation cannot generally be integrated farther, but in this and every similar case the integration can be performed if we suppose x always very small. Suppose the particle then to have been at rest at N, and to have been slightly displaced from this position of equilibrium, the displacement being so small that throughout the motion $\frac{x^2}{b^2}$ may be neglected in comparison with $\frac{x}{b}$. We have then from (1),

$$\frac{d^2x}{dt^2} = -\frac{\mu x}{b^3}\left(1+\frac{x^2}{b^2}\right)^{-\frac{3}{2}}$$

$$= -\frac{\mu x}{b^3}\left(1 - \frac{3}{2}\frac{x^2}{b^2} + \&c.\right)$$

$$= -\frac{\mu x}{b^3} \text{ nearly};$$

$$\text{or } \frac{d^2x}{dt^2} + \frac{\mu x}{b^3} = 0,$$

the same equation of motion as in Art. (42). The motion will therefore be oscillatory, the time of each small oscillation being $2\pi\sqrt{\frac{b^3}{\mu}}$.

54. *A particle moves in a straight line under the action of a force in that line varying inversely as the nth power of the distance of the particle from a fixed point; to determine the motion.*

Measuring x as before, the equation of motion will be

$$\frac{d^2x}{dt^2} = -\frac{\mu}{x^n}.$$

Multiplying by $2\frac{dx}{dt}$ and integrating

$$\left(\frac{dx}{dt}\right)^2 = v^2 = C + \frac{2}{n-1} \cdot \frac{\mu}{x^{n-1}}.$$

Suppose the particle to start from rest at a distance a, then when $x = a$, $v = 0$; $\therefore\ C = -\frac{2}{n-1} \cdot \frac{\mu}{a^{n-1}}$;

$$\therefore\ \left(\frac{dx}{dt}\right)^2 = v^2 = \frac{2\mu}{n-1}\left(\frac{1}{x^{n-1}} - \frac{1}{a^{n-1}}\right) \text{ ...... (1).}$$

55. This equation cannot generally be integrated farther, but if we suppose the particle to have started from a point at an infinite distance, we have $a = \infty$, and

$$v^2 = \frac{2\mu}{n-1} \cdot \frac{1}{x^{n-1}},$$

where v is the velocity from infinity, at the distance x.

We have also in this case

$$\frac{dx}{dt} = \left(\frac{2\mu}{n-1}\right)^{\frac{1}{2}} \frac{1}{x^{\frac{n-1}{2}}};$$

$$\therefore\ \frac{dt}{dx} = \left(\frac{n-1}{2\mu}\right)^{\frac{1}{2}} x^{\frac{n-1}{2}}.$$

Integrating this between the limits $x = \alpha$, $x = \beta$, we have for the time of moving from $x = \alpha$ to $x = \beta$;

$$T = \frac{2}{n+1} \cdot \left(\frac{n-1}{2\mu}\right)^{\frac{1}{2}} (\alpha^{\frac{n+1}{2}} - \beta^{\frac{n+1}{2}}).$$

56. If we expand (1) into a series, we obtain

$$\frac{dt}{dx} = -\left(\frac{n-1}{2\mu}\right)^{\frac{1}{2}} x^{\frac{n-1}{2}} \left(1 - \frac{x^{n-1}}{a^{n-1}}\right)^{-\frac{1}{2}}$$

$$= -\left(\frac{n-1}{2\mu}\right)^{\frac{1}{2}} x^{\frac{n-1}{2}} \{1 + \frac{1}{2} \cdot \frac{x^{n-1}}{a^{n-1}} + \frac{1\,.\,3}{2\,.\,4} \cdot \frac{x^{2n-2}}{a^{2n-2}} + \&c.\}.$$

Integrating between the limits $x = a$, $x = 0$, we get for the time of falling to the center from a distance a the expression

$$\left(\frac{n-1}{2\mu}\right)^{\frac{1}{2}} 2a^{\frac{n+1}{2}} \left\{\frac{1}{n+1} + \frac{1}{2} \cdot \frac{1}{3n-1} + \&c.\right\},$$

which therefore for different distances varies as $a^{\frac{n+1}{2}}$.

57. The above solution fails when $n = 1$, but the time of falling to the center may be found as follows. The expression in Art. (47), gives

$$\left(\frac{dx}{dt}\right)^2 = C - 2\mu \log x$$

$$= 2\mu \log \frac{a}{x},$$

since when $x = a$, $\frac{dx}{dt} = 0$. Hence,

$$\sqrt{(2\mu)} \frac{dt}{dx} = -\frac{1}{\sqrt{\left\{\log\left(\frac{a}{x}\right)\right\}}},$$

the negative sign being taken since x diminishes as t increases. Put T for the required time, then

$$\sqrt{(2\mu)}\, T = -\int_a^0 \frac{dx}{\sqrt{\left\{\log\left(\frac{a}{x}\right)\right\}}}.$$

To transform the integral put $\sqrt{\left\{\log\left(\frac{a}{x}\right)\right\}} = y$. Then we have

$$x = a\epsilon^{-y^2}, \text{ and } \frac{dx}{dy} = -2a\epsilon^{-y^2}y,$$

and the limits of y are 0 and ∞. Hence

$$\sqrt{(2\mu)} \cdot T = 2a\int_0^\infty \epsilon^{-y^2} dy,$$

which (Gregory's Examples, p. 466)

$$= 2a \cdot \tfrac{1}{2}\sqrt{(\pi)};$$

$$\therefore\ T = a\sqrt{\left(\frac{\pi}{2\mu}\right)},$$

and is therefore directly as the space traversed.

58. *A particle is constrained to move in a straight line, and is acted on by a force directed to a point not in that line, and expressed by a function $\phi(r)$ of the distance; to determine the time of a small oscillation.*

Take the same notation as in Art. (52), then the force along PO being $\phi(r)$ its component along PN is $\phi(r)\frac{x}{r}$; therefore the equation of motion is

$$\frac{d^2x}{dt^2} = -\phi(r)\frac{x}{r}.$$

But $r = \sqrt{(b^2 + x^2)} = b\sqrt{\left(1 + \frac{x^2}{b^2}\right)}$

$$= b \text{ approximately};$$

$$\therefore\ \frac{d^2x}{dt^2} + \frac{\phi(b)}{b} \cdot x = 0,$$

and therefore by Art. (42), the time of a small oscillation

$$= 2\pi \sqrt{\left\{\frac{b}{\phi(b)}\right\}}.$$

EXAMPLES.

(1) A body is projected vertically upwards with a velocity which will carry it to a height $2g$ feet; shew that after three seconds it will be descending with a velocity g.

(2) The straight line down which a particle will slide in the shortest time from a given point to a given circle in the same vertical plane, is the line joining the point to the upper or lower extremity of the vertical diameter, according as the point is within or without the circle.

(3) Find the line of quickest descent from the focus to a parabola whose axis is vertical and vertex upwards, and shew that its length is equal to the latus rectum.

(4) The locus of all points in the same vertical plane for which the least time of sliding down an inclined plane to a circle is constant is another circle.

(5) Two bodies fall in the same time from two given points in space in the same vertical down two straight lines drawn to any point of a surface, shew that the surface is an equilateral hyperboloid of revolution, having the given points as vertices.

(6) A semicycloid is placed with its axis vertical and vertex downwards, and from different points in it a number of particles are let fall at the same instant, each moving down the tangent at the point from which it sets out; prove that they will reach the involute (which passes through the vertex) all at the same instant.

(7) A particle moves in a straight line under the action of a force varying inversely as the $\left(\frac{3}{2}\right)^{\text{th}}$ power of the distance, shew that the velocity acquired by falling from an infinite

distance to a distance a from the center is equal to the velocity which would be acquired in moving from rest at a distance a to a distance $\frac{a}{4}$.

(8) A particle moves in a straight line from a distance a towards a center of force, the force varying inversely as the cube of the distance; shew that the whole time of descent

$$= \frac{a^2}{\sqrt{(\mu)}}.$$

(9) A particle is placed at a given point between two centers of force of equal intensity attracting directly as the distance; to determine the motion and the time of an oscillation.

Let $2a$ be the distance between the centers, x the distance of the particle at any time from the middle point between them, then the equation of motion is

$$\frac{d^2x}{dt^2} = -\mu(a+x) + \mu(a-x)$$

$$= -2\mu x.$$

Hence, the time of an oscillation $= \frac{\pi}{\sqrt{(2\mu)}}$.

(10) A particle acted upon by two centers of force, each attracting with an intensity varying inversely as the square of the distance, is projected from a given point between them, to find the velocity of projection that the particle may just arrive at the neutral point of attraction and remain at rest there.

Taking one of the points as origin, and supposing $2a_1$, $2a_2$, to be the original distances of the particle from the centers, the equation of motion will be

$$\frac{d^2x}{dt^2} = \frac{\mu}{(a_1+a_2-x)^2} - \frac{\mu'}{(a_1+a_2+x)^2}.$$

This is to be integrated as usual, and then V determined so that when the particle arrives at the neutral point $v=0$.

This gives

$$V^2 = \frac{\{\sqrt{(\mu)}\,a_2 - \sqrt{(\mu')}\,a_1\}^2}{a_1 a_2 (a_1 + a_2)}.$$

(11) Supposing the Earth a homogeneous spheroid of equilibrium, the time of descent of a body let fall from any point P on the surface down a hole bored to the center C, varies as CP, and the velocity at the center is constant.

(12) A material particle placed at a center of attraction varying as the distance, is urged from rest by a constant force which acts for one-sixth of the time of a complete oscillation about the center, ceases for the same period, and then acts as before, shew that the particle will then be retained at rest, and that the spaces moved through in the two periods are equal.

(13) A body moves from rest at a distance a towards a center of force, the force varying inversely as the distance: shew that the time of describing the space between βa and $\beta^n a$ will be a maximum if $\beta = \dfrac{1}{n^{\frac{1}{2(n-1)}}}$.

(14) If the time of a body's descent in a straight line towards a given center of force vary inversely as the square of the distance fallen through, determine the law of the force.

(15) Assuming the velocity of a body falling to a center of force to be as $\sqrt{\left(\dfrac{a-x}{x}\right)}$, where a is the initial and x the variable distance from the center, find the law of the force.

(16) Find the time of falling to the center when the force $\propto (\text{dist.})^{-\frac{5}{3}}$.

(17) Shew that the time of descent, to a center of force $\propto (\text{dist.})^{-2}$, through the first half of the initial distance, is to that through the last half as $\pi + 2 : \pi - 2$.

(18) A particle descends to a center of force $\propto (\text{dist.})^n$. Find n so that the velocity acquired from infinity to distance

a, shall be equal to that acquired from distance a to distance $\frac{1}{4}a$, from the center.

(19) A particle is placed at the extremity of the axis of a thin attracting cylinder of indefinite length and of radius a, shew that its velocity after describing a space x is proportional to

$$\sqrt{\left\{\log . \frac{x + \sqrt{(x^2 + a^2)}}{a}\right\}}.$$

(20) A particle falls to an infinite solid bounded by a plane face, find the time of descent.

(21) Every point of a fine uniform ring repels with a force $\propto$ (dist.)$^{-2}$, find the time of a small oscillation in its plane, about the center.

(22) Shew that a body cannot move so that the velocity shall vary as the space from the beginning of the motion. And if the velocity vary as the cube root of that space, determine the force, and the time of describing a given space.

(23) Shew that the time of quickest descent down a focal chord of a parabola whose axis is vertical is

$$\sqrt{\left(\frac{3^{\frac{3}{2}} l}{g}\right)}.$$

where l is the latus rectum.

(24) An ellipse is suspended with its major axis vertical, find the diameter down which a particle will fall in the least time, and the limiting value of the excentricity that this may not be the axis major itself.

(25) Particles slide down chords from a point O to a curved surface, under the action of a plane whose attraction is as the distance, and they reach the surface in the same time; shew that the surface is generated by the revolution (about a line whose length is a through O perpendicular to the plane) of the curve whose polar equation about O is

$$\rho \cos \theta = a \{1 - \cos (k \cos \theta)\}.$$

(26) If the particles commence their motion at the surface, and reach O after a given time, the equation to the generating curve is

$$\rho \cos\theta = a\{\sec(k\cos\theta) - 1\}.$$

(27) Prove that the times of falling through a given space AC towards a center of force S, under the action of two forces, one of which varies as the distance, and the other is constant and equal to the original value of the first, are as the arc (whose versed sine is AC) to the chord, in a circle whose radius is AS.

(28) The earth being supposed a thin spherical shell, in the surface of which a circular aperture is made, if a particle be dropped from the center of the aperture, determine its velocity at any point of the descent.

(29) If a particle fall down a radius of a circle under the action of a force $\propto (D)^3$ in the center, and ascend the opposite radius under the action of the same force supposed repulsive, shew that it will acquire a velocity which is a geometric mean between radius, and the force at the circumference.

(30) If a particle fall to a center of force $\propto (D)$; determine the constant force which would produce the effect in the same time, and compare the final velocities.

(31) Find the equation to the curve down each of whose tangents a particle will slide to the horizontal axis in a given time.

(32) A sphere is composed of an infinite number of free particles equally distributed which gravitate to each other without interfering; supposing the particles to have no initial velocity, prove that the mean density about a given particle will vary inversely as the cube of its distance from the center.

CHAPTER IV.

PARABOLIC MOTION.

59. In this chapter we intend to treat principally of the motion of a free particle which is subject to the action of forces whose resultant is parallel to a given fixed line.

The simplest case of course will be when that resultant is constant. The problem then becomes the determination of the motion of a projectile in vacuo, as the attraction of the earth may be considered within moderate limits as constant and parallel to a fixed line. This we will now consider.

60. Taking the axis of x horizontal and in the vertical plane and sense of projection, and that of y vertically upwards, it is evident that the particle will continue to move in the plane of xy, as it is projected in it, and is subject to no force which would tend to withdraw it from that plane.

The equations of motion then are

$$m\,.\frac{d^2x}{dt^2}=0,\quad m\,.\frac{d^2y}{dt^2}=-mg.$$

Suppose that the point from which the particle is projected is taken as origin, that the velocity of projection is V, and that the direction of projection makes an angle α with the axis of x.

The first and second integrals of the above equations will then be

$$\frac{dx}{dt}=V\cos\alpha,\quad \frac{dy}{dt}=V\sin\alpha-gt,$$

$$x=V\cos\alpha\,.\,t,\quad y=V\sin\alpha\,.\,t-\tfrac{1}{2}gt^2\ldots\ldots(1).$$

These equations give the co-ordinates of the particle and its velocity parallel to either axis for any assumed value of the time.

Eliminating t between equations (1) we obtain the equation to the trajectory, viz.

$$y = x \tan\alpha - \frac{g}{2V^2\cos^2\alpha} \, . \, x^2 \ldots\ldots\ldots\ldots (2),$$

which shews that the particle will move in a parabola whose axis is vertical, and vertex upwards.

61. Equation (2) may be written

$$x^2 - \frac{2V^2 \sin\alpha\cos\alpha}{g} \, . \, x = -\frac{2V^2\cos^2\alpha}{g} \, . \, y,$$

$$\text{or } \left(x - \frac{V^2\sin\alpha\cos\alpha}{g}\right)^2 = -\frac{2V^2\cos^2\alpha}{g}\left(y - \frac{V^2\sin^2\alpha}{2g}\right).$$

And by comparing this with the equation to a parabola, we find for the co-ordinates $x_0 y_0$ of the vertex

$$x_0 = \frac{V^2\sin\alpha\cos\alpha}{g}, \quad y_0 = \frac{V^2\sin^2\alpha}{2g}.$$

Hence we obtain the equation to the directrix

$$y = y_0 + \tfrac{1}{4}\,(\text{parameter}) = \frac{V^2\sin^2\alpha}{2g} + \frac{V^2\cos^2\alpha}{2g} = \frac{V^2}{2g}.$$

Now if v be the velocity of the particle at any point of its path

$$\begin{aligned} v^2 &= \overline{\frac{dx}{dt}}\Big|^2 + \overline{\frac{dy}{dt}}\Big|^2 \\ &= (V^2\cos^2\alpha) + (V^2\sin^2\alpha - 2Vg\sin\alpha \, . \, t + g^2t^2) \\ &= V^2 - 2g\,(V\sin\alpha \, . \, t - \tfrac{1}{2}gt^2) \\ &= V^2 - 2gy, \end{aligned}$$

which is that due to a fall from rest through a height $\frac{v^2}{2g}$, or $\frac{V^2}{2g} - y$, i.e. through the distance from the directrix.

62. *To find the time of flight along a horizontal plane.*

Put $y=0$ in equation (2). The corresponding values of x are 0 and $\frac{2V^2}{g}\sin\alpha\cos\alpha$. But the horizontal velocity is $V\cos\alpha$. Hence the time of flight is $\frac{2V\sin\alpha}{g}$, and ceteris paribus varies as the sine of the inclination to the horizon of the direction of projection.

To find the time of flight along a plane through the origin whose intersection with the plane of projection makes an angle β with the horizon, we must eliminate y between (2) and $y = x\tan\beta$.

This gives for the abscissa of the point where the projectile meets the plane,

$$x_1 = \frac{2V^2}{g}(\sin\alpha\cos\alpha - \tan\beta\cos^2\alpha)$$

$$= \frac{2V^2\cos\alpha\sin(\alpha-\beta)}{g\cos\beta}.$$

Hence time of flight

$$= \frac{x_1}{V\cos\alpha} = \frac{2V}{g}\cdot\frac{\sin(\alpha-\beta)}{\cos\beta}.$$

63. The range on the horizontal plane is $\frac{V^2}{g}\sin 2\alpha$. For a given value of V this will be greatest when

$$2\alpha = \frac{\pi}{2}, \text{ or } \alpha = \frac{\pi}{4}.$$

That on the inclined plane is $\frac{x_1}{\cos\beta}$, or

$$\frac{2V^2}{g\cos^2\beta}\cdot\cos\alpha\sin(\alpha-\beta).$$

That this may be a maximum for a given value of V we must equate to zero its differential coefficient with respect to α, which gives the equation

$$\cos\alpha\cos(\alpha-\beta)-\sin\alpha\sin(\alpha-\beta)=0,$$

$$\text{or } \cos(2\alpha-\beta)=0;$$

$$\therefore\ \alpha=\frac{1}{2}\left(\frac{\pi}{2}+\beta\right).$$

Hence the direction of projection required for the greatest range makes with the vertical an angle

$$=\frac{\pi}{2}-\alpha=\frac{1}{2}\left(\frac{\pi}{2}-\beta\right),$$

or bisects the angle between the vertical and the plane on which the range is measured.

64. *To find the elevation necessary to the particle's passing through a given point.*

Suppose the point in the axis of x and distant a from the origin. Then we must have

$$\frac{V^2}{g}\sin 2\alpha=a.$$

$$\text{Let } \sin^{-1}\frac{ag}{V^2}=\alpha'.$$

Then the values of α will be $\frac{\alpha'}{2}$ and $\frac{\pi-\alpha'}{2}$, so that we see there are two directions in which a particle may be projected so as to reach the given point, and that these are equally inclined to the direction of projection $\left(\alpha=\frac{\pi}{4}\right)$ which gives the greatest range.

Suppose the given point in the plane which makes an angle β with the horizon. Then if its abscissa be a, we must have

$$\frac{2V^2}{g\cos\beta}\cdot\cos\alpha\sin(\alpha-\beta)=a.$$

If α', α'' be the two values of α which satisfy this equation, we must have

$$\cos\alpha'\sin(\alpha'-\beta)=\cos\alpha''\sin(\alpha''-\beta);$$

$$\text{and therefore } \alpha'' - \beta = \frac{\pi}{2} - \alpha',$$

$$\text{or } \alpha'' - \frac{1}{2}\left(\frac{\pi}{2} + \beta\right) = \frac{1}{2}\left(\frac{\pi}{2} + \beta\right) - \alpha'.$$

Hence as before the two directions of projection which enable the particle to strike a point in a given plane through the point of projection are equally inclined to the direction of projection required for the greatest range along that plane.

65. *To find the envelope of all the trajectories corresponding to different values of* α.

Differentiating equation (2) with respect to α, we get

$$\sec^2 \alpha - \frac{gx}{V^2} \cdot \frac{\sin \alpha}{\cos^3 \alpha} = 0,$$

$$\text{or } \tan \alpha = \frac{V^2}{gx} \quad \text{.................} \quad (3).$$

The elimination of α between (2) and (3) gives us as the equation to the required envelope

$$y = \frac{V^2}{2g} - \frac{gx^2}{2V^2},$$

$$\text{or } x^2 = -\frac{2V^2}{g}\left(y - \frac{V^2}{2g}\right);$$

which is evidently a parabola whose axis is vertical, whose focus is the point of projection, and whose vertex is in the common directrix of the trajectories.

It will easily be seen from what has gone before that there are two directions of projection, so that the particle may pass through any given point within this parabola, only one for a point in it, and of course there is no possibility of its reaching (with the given velocity V) any point without this parabola.

66. By a somewhat simpler method of considering the problem we might easily have arrived at some of the more obvious properties of the trajectory, thus.

Take the direction of projection as the axis of x, and the vertical downwards from the point of projection as that of y. By the second law of motion we may consider the velocity due to projection to be maintained constant $= V$ parallel to the axis of x, while we have in addition parallel to the axis of y the portion due to gravity as investigated in Chap. III.

$$\text{Hence } \left.\begin{aligned} x &= Vt \\ y &= \tfrac{1}{2}gt^2 \end{aligned}\right\} \text{ at any time,}$$

$$\text{and } \therefore\ x^2 = \frac{2V^2}{g}y,$$

the equation to a parabola referred to a diameter and the tangent at its vertex. The distance of the origin from the directrix, being $\frac{1}{4}$th of the coefficient of y, is $\frac{V^2}{2g}$, and the velocity due to a fall through that space is as before

$$\sqrt{\left(2g \cdot \frac{V^2}{2g}\right)} = V.$$

67. Many properties of parabolic motion are more easily obtained by geometry than by analysis. We proceed to give a few examples.

Thus in the figure suppose O to be the point of projection, MN the directrix common to the trajectories of all particles projected from O with a given velocity, and suppose it be required to determine the direction of projection for the greatest range along the plane OS. Since O is a point in each trajectory and MN the common directrix, all the foci will lie in the circle $MF'FF''$ described with center O and touching MN in M.

Take any point in this circle as F', then the path whose focus is F' will intersect OS again in a point P' such that if $P'N'$ be drawn perpendicular to MN, $F'P' = P'N'$. Now in order that P' may be as far as possible from O, at P suppose, it is evident ex absurdo that the focus must be taken at the point F where OS meets the circle. But the tangent at O bisects the angle between the diameter MO and the focal distance OF. Hence the direction of projection for the greatest range on an inclined plane bisects the angle between the plane and the vertical.

Again, if with center P' and radius $P'F'$ an arc be described cutting $F'FF'''$ in F'', it is evident that the trajectories

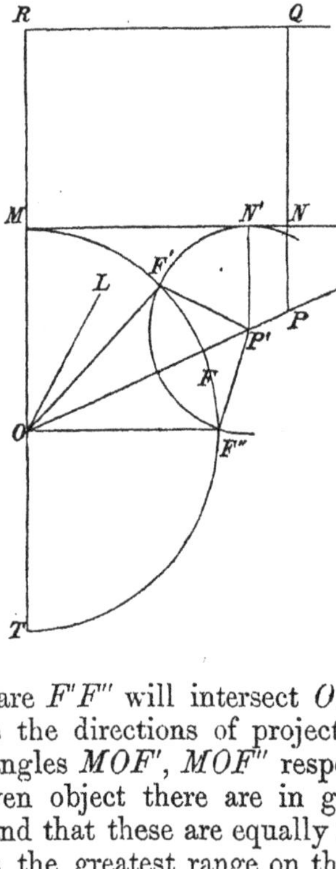

whose foci are $F'F''$ will intersect OS in the same point P'. Hence since the directions of projection for these cases will bisect the angles MOF', MOF'' respectively, we see that to strike a given object there are in general two directions of projection, and that these are equally inclined to the direction which gives the greatest range on the plane passing through the object and the point of projection.

Again, for the envelope of all the trajectories. It is evident that P must be a point in the envelope since it is a point in one of the parabolas, and there can be no point in any of them along the line OS farther from O than P is. Draw PN perpendicular to MN, and produce it till $NQ = FO$. Draw QR parallel to MN, and cutting OM in R. RQ is a fixed line since $RM = MO$, and as $OP = PQ$ we see that the envelope is a parabola whose focus is O and directrix RQ.

It may be seen at once that it touches in P the only trajectory which can pass through that point. Since the tangents of both at P bisect the angle OPQ or FPN.

68. Ex. *It is required to throw a shell with given velocity so as to strike at right angles an inclined plane through the point of projection.*

The letters being the same as before, join ST cutting

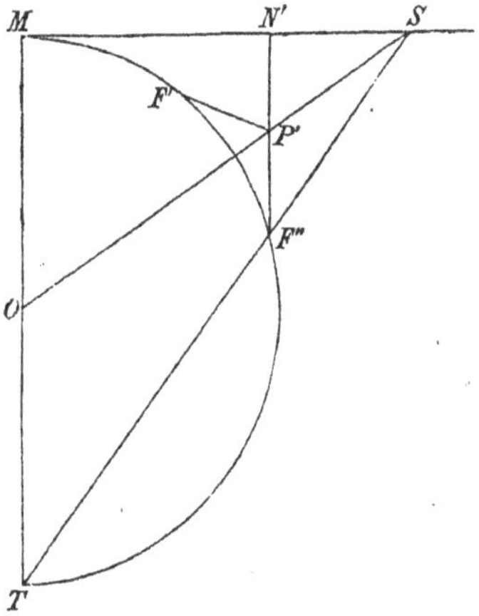

$MF'F''$ in F''. Draw $F''P'N'$ perpendicular to MS cutting OS in P'. Find F' so that $P'F' = P'F'' = P'N'$. P' is a point in the trajectory whose focus is F'. Hence the tangent at P' bisects $F'P'N'$. But OP' bisects $F'P'F''$. Hence the trajectory at P' is perpendicular to OS.

Also as F'' is the focus of the other path by which the point P' might be reached, P' will be the vertex of that path, and therefore the particle will be moving horizontally when it reaches P'.

69. Even if the plane along which the range is measured do not pass through the point of projection, a somewhat

similar construction will enable us to find the direction of projection for a maximum. Thus,

Let it be required to find the direction of projection from

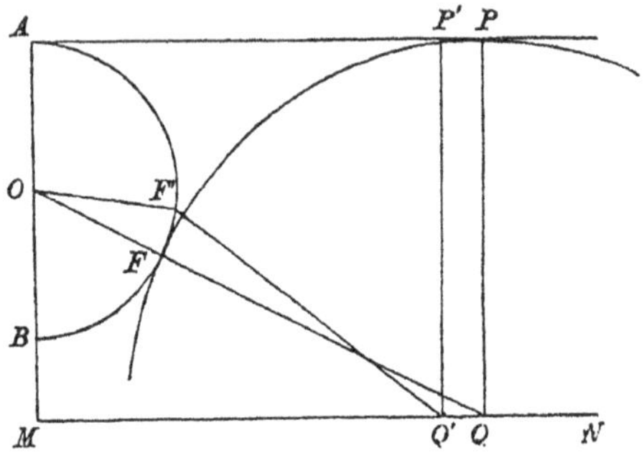

O with velocity due to AO in order that the range on a horizontal line MN may be a maximum.

Suppose Q' the point where the projectile falls. Join $Q'F''$, $F''O$, F'' being the focus of the path. Then if $Q'P'$ be vertical and meet the horizontal line through A in P', we have $F''Q' = Q'P'$. This is true of each of the paths, and $Q'P'$ is constant. The farthest point Q which can be reached will therefore be determined by inflecting OQ to MN, where $OQ = OA + PQ$, and therefore if $AO = a$, $AM = b$, the cosine of double the requisite angle of elevation will be $\left(\frac{b-a}{b+a}\right)$.

Should MN be an inclined plane, we must evidently draw a line QO, and the corresponding vertical QP; such that if QO meet the circle in F, $FQ = QP$.

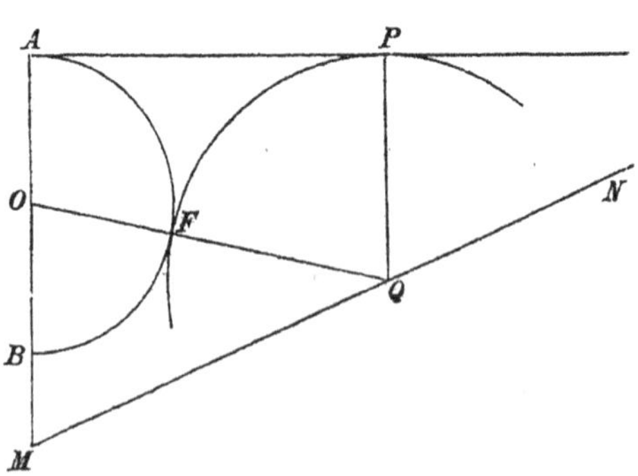

This resolves itself into the well-known geometrical problem of describing a circle whose center is in a given line, and which touches a given circle, and a given straight line.

70. Perhaps however the most satisfactory method of solving all such problems about the maximum range, is to describe the parabola which envelopes all the trajectories. The point where this cuts the plane, &c. on which the range is estimated, gives the maximum value of the latter, and it is then easy from known properties of the envelope to construct for the required path.

71. Let P be any point in the trajectory, S its focus, BN, AL, the directrix, and the tangent at the vertex.

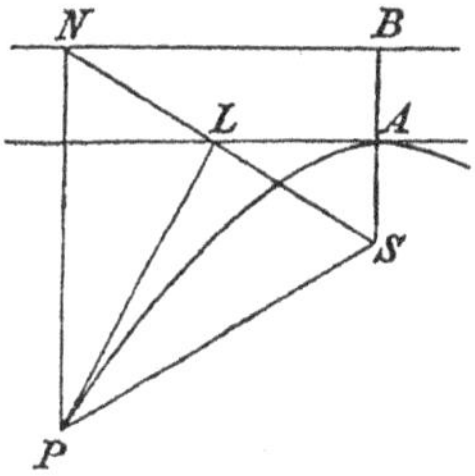

Then $(\text{velocity at } P)^2 = 2g \,.\, PN = 2g \,.\, SP$

$$= (\text{by a property of the parabola}) \frac{2g}{SA} \,.\, SL^2 = \frac{g}{2SA} \,.\, SN^2.$$

Hence velocity at $P \propto SN$, and since by the figure $SL = LN$; $\therefore PL$ is the tangent at P and is perpendicular to SN.

Hence as SN is perpendicular to the direction of motion at P, proportional to the velocity at P, and drawn from a fixed point S, N is a point in the Hodograph (see Chap. IX.) which is thus shewn to be a horizontal straight line.

This proposition will be found of considerable utility in solving various problems in projectiles. It is evident that SB, BN represent the horizontal and vertical velocities at P, in the same scale in which SN represents the entire velocity at that point.

72. *When a particle moves subject to the action of two centers of force where the law is the direct distance and the absolute intensities the same, but one attractive and the other repulsive, its motion will be the same as that of a projectile.*

For the whole force on the particle resolved perpendicular to the line joining the centers is evidently zero, and that parallel to this line is equal to that which would be exerted by either of the centers on a particle placed at the other; and always tends in the direction from the repelling, to the attracting, center. It corresponds therefore exactly to the force of gravity.

73. Again if a particle moves on a plane inclined to the horizon at an angle θ, the whole force on it will be $g \sin \theta$ parallel to the line of greatest slope on the plane, and therefore the trajectory will still be a parabola, whose dimensions will depend upon θ.

Ex. *A particle is projected horizontally from a given point with a given velocity, and moves on an inclined plane; find the locus of the directrices of its path for different inclinations of the plane.*

It will be easily seen that when a particle moves on an inclined plane, the velocity at any point is equal to that which would have been acquired by sliding from the directrix; that is (39) equal to the velocity due to the fall from a horizontal plane through the directrix. Now the velocity is given constant, hence the locus of the directrices is a horizontal plane.

74. Suppose now a particle moving subject to the action of a force always perpendicular to a given plane, and a function of the distance of the particle from the plane.

It is evident that the motion will be confined entirely to a plane through the direction of projection perpendicular to the attracting plane. Let us take the former as the plane of xy, the axis of x lying in the attracting plane. Let $\phi'(D)$ be the accelerating force at distance D. Then the equations of motion are

$$\frac{d^2x}{dt^2} = 0, \quad \frac{d^2y}{dt^2} = -\phi'(y).$$

Suppose the particle projected from a point (a, b) in a direction making an angle α with the axis of x, and with a velocity V.

Multiplying by $2\frac{dx}{dt}$, $2\frac{dy}{dt}$, and integrating we get

$$\left.\begin{aligned}\overline{\frac{dx}{dt}}\bigg|^2 &= \text{const.} = V^2\cos^2\alpha,\\ \overline{\frac{dy}{dt}}\bigg|^2 &= C - 2\phi(y) = V^2\sin^2\alpha + 2\phi(b) - 2\phi(y),\end{aligned}\right\}\ \ldots\ldots (1).$$

Hence $v^2 = V^2 + 2\{\phi(b) - \phi(y)\}$ and therefore depends only on the distance from the attracting plane.

To find the differential equation to the path, we have

$$\frac{\frac{dy}{dt}}{\frac{dx}{dt}} = \frac{dy}{dx} = \frac{\sqrt{[V^2\sin^2\alpha + 2\{\phi(b) - \phi(y)\}]}}{V\cos\alpha},$$

an equation only integrable for particular forms of the function ϕ. The only interesting case is that in which the attraction of the plane is inversely as the cube of the distance,

$$\text{or } \phi'(y) = \frac{\mu}{y^3}, \text{ and therefore } \phi(y) = -\tfrac{1}{2}\cdot\frac{\mu}{y^2}.$$

The differential equation becomes

$$\frac{dy}{dx} = \frac{\sqrt{\left\{\left(V^2\sin^2\alpha - \frac{\mu}{b^2}\right) + \frac{\mu}{y^2}\right\}}}{V\cos\alpha}.$$

There will be three cases according as $\left(V^2\sin^2\alpha - \frac{\mu}{b^2}\right)$ is positive, zero or negative.

1st. Let it be positive and $= \left(\frac{\mu}{a_1^2}\right)$,

$$y\cdot\frac{dy}{dx} = \frac{\sqrt{(\mu)}}{Va_1\cos\alpha}\sqrt{(a_1^2 + y^2)};$$

$$\therefore \sqrt{(a_1^2 + y^2)} = \frac{\sqrt{(\mu)}}{Va_1 \cos \alpha} (x + C),$$

the equation to an hyperbola one of whose axes is that of x.

2nd. Let $V^2 \sin^2 \alpha - \frac{\mu}{b^2} = 0$,

$$y \cdot \frac{dy}{dx} = \frac{\sqrt{(\mu)}}{V \cos \alpha},$$

$$\text{and } y^2 = \frac{2\sqrt{(\mu)}}{V \cos \alpha} (x + C),$$

a parabola whose axis is that of x.

3rd. Let $V^2 \sin^2 \alpha - \frac{\mu}{b^2}$ be negative, and $= -\frac{\mu}{a_1^2}$,

$$y \cdot \frac{dy}{dx} = \frac{\sqrt{(\mu)}}{Va_1 \cos \alpha} \sqrt{(a_1^2 - y^2)},$$

$$\text{or } \sqrt{(a_1^2 - y^2)} = \frac{\sqrt{(\mu)}}{Va_1 \cos \alpha} (x + C),$$

the equation to an ellipse of which the axis of x is an axis.

We might have obtained the above results by integrating separately the two equations of motion, and then eliminating t between them.

From this we see that the conic sections are the only curves which can be described by a free particle moving in a plane and subject to a force in the direction, and inversely as the cube, of the perpendicular on a given line in that plane.

The converse is easily investigated, thus:

75. *To find the law of force perpendicular to the axis that a free particle may describe a conic section.*

Take the axis as that of x, then the equation

$$y^2 = 2mx + nx^2, \ (1),$$

will represent by properly taking m and n, any given conic.

Also since the force is perpendicular to the axis

$$\frac{dx}{dt} = c. \quad \text{Hence}$$

$$y \cdot \frac{dy}{dt} = mc + nxc;$$

$$\therefore\ y \cdot \frac{d^2y}{dt^2} + \left(\frac{dy}{dt}\right)^2 = nc^2;$$

$$\therefore\ \frac{d^2y}{dt^2} = \frac{1}{y}\left\{nc^2 - \left(\frac{dy}{dt}\right)^2\right\}$$

$$= \frac{1}{y}\left\{nc^2 - \frac{(m+nx)^2}{y^2}c^2\right\}$$

$$= \frac{c^2}{y^3}(ny^2 - m^2 - 2mnx - n^2x^2)$$

$$= -\frac{c^2m^2}{y^3} \text{ by equation (1).}$$

76. *To find the force which must act perpendicular to a given plane, that a given path may be described.*

Take the axes as before; then Y being the required force (a function of y only) we have

$$\frac{d^2x}{dt^2} = 0, \text{ or } \frac{dx}{dt} = \text{const.} = a, \text{ suppose}$$

$$\frac{d^2y}{dt^2} = Y, \ldots\ldots\ldots\ldots\ldots (1).$$

Let $y = f(x)$ be the equation to the given curve

$$\frac{dy}{dt} = af'(x),$$

$$\frac{d^2y}{dt^2} = a^2f''(x),$$

$$\text{or by (1),}\quad Y = a^2 f''(x)$$
$$= a^2 f''\{f^{-1}(y)\},$$

by the equation to the curve. Hence the law of force required is found.

77. It is necessary to observe that in all these cases, when the particle actually reaches the axis, it will not proceed to describe the portion of the same curve which lies on the

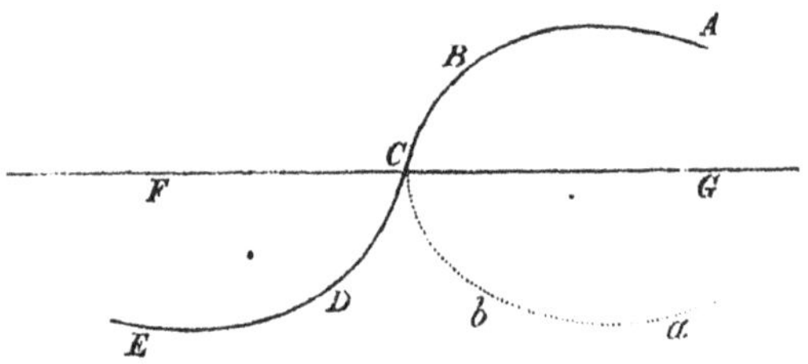

other side of the axis, as this would involve a change in sign of the constant horizontal velocity. It is evident that in such cases the particle having described ABC will, instead of pursuing the course Cba, actually describe CDE similar and equal to Cba, but turned in the opposite direction.

Although in the case of ABC being a conic one of whose axes is CG, and therefore cutting it at right angles in C it might seem that at C the horizontal velocity vanishes, yet it is to be recollected that the velocity at C is indefinitely great, and it may easily be shewn by independent methods, such as limits, if the foregoing analysis do not appear satisfactory, that the velocity parallel to CG is really constant throughout the motion.

78. It may be useful to notice that cases of this kind are reduced at once to investigations similar to those of last Chapter, by considering separately the equations of motion, parallel and perpendicular to the attracting plane.

Whenever then we can completely determine the motion of a particle towards a center of force, in a straight line, we can also completely solve the problem of the motion of a particle anyhow projected, and attracted by an indefinite plane; the law of force in terms of the distance being the same in the two cases.

79. Generally, when a particle is anyhow projected and subject only to the action of a force whose direction is parallel to a given line; the velocity perpendicular to that line is constant, and in passing from any point to another, the square of the entire velocity is altered by a quantity depending only upon the position of the projections of these two points on the given line.

Take the axis of y parallel to the given line and the axis of x at right angles to it, so that the direction of projection lies in xy. This will evidently be the plane of motion; and the equations are

$$\frac{d^2x}{dt^2} = 0,$$

$$\frac{d^2y}{dt^2} = Y;$$

$$\therefore \frac{dx}{dt} = c,$$

$$\text{and } v^2 = \left.\frac{dx}{dt}\right|^2 + \left.\frac{dy}{dt}\right|^2 = V^2 + 2\int_{y_1}^{y} Y dy$$

$$= V^2 + \phi(y_1 y),$$

V being the velocity of projection, and y_1 the co-ordinate of the point of projection; which proves the proposition.

80. As another example of the motion of a particle under the action of forces whose direction is constant, let us consider the motion of a particle of light in the emission theory, at the confines of two media whose bounding surface is plane.

In this case the hypothesis is that the attractive or repulsive forces exerted by the particles of any medium on a particle of light passing through it, are insensible at sensible distances but enormously great at indefinitely small distances. Hence of course the path of such a particle in an uniform medium will be a straight line, and will be described with constant velocity, until the particle is indefinitely near to the bounding surface of the medium.

Thus, suppose AB to be the common plane surface of two media. Draw CD at a distance from AB equal to that at which the intensity of the attractive forces of the particles of the medium begins to be sensible; and draw EF parallel to CD and equidistant from it with AB. By what we have just noticed, a particle of light moving along PQ will arrive at Q without any change of velocity or direction. Also from the symmetry of the figure, the resultant of all the sensible forces on it will always be perpendicular to AB. This shews (79) that the velocity resolved parallel to AB is constant throughout the motion, and also that whatever be the direction of PQ, the change in the square of the velocity in passing from Q to any point of the path will depend only on the distance of that point from AB.

Let PQR represent a portion of the path.

We have no means of determining its actual form, since the extent in which the force is sensible, the law of its variation, and whether it change from attractive to repulsive with the distance, are unknown.

Through any point R draw KRL parallel to AB, and let GH be equidistant from KL with AB.

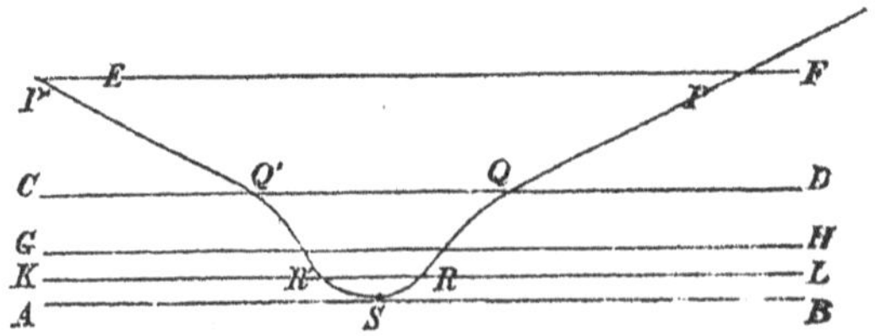

Then at R the particle is subject only to the actions of the upper medium beyond GH, and of the lower medium.

If the resultant effects of these two should at a point S in the superior medium destroy the velocity perpendicular to AB, the particle will evidently pursue a course $SR'Q'P'$ similar and equal to $SRQP$, and the angles $P'Q'C$ and PQD will be equal as also the velocities in PQ and $P'Q'$. (Art. 79). Hence we have the case of a ray reflected at a plane surface.

If however the attraction of the lower medium should so prevail that the particle actually enters it, then we may consider its motion while still within the range of action of both media precisely as before; but there will be two cases.

I. At some point as S whose distance from AB is less than that of AB from CD, the velocity perpendicular to

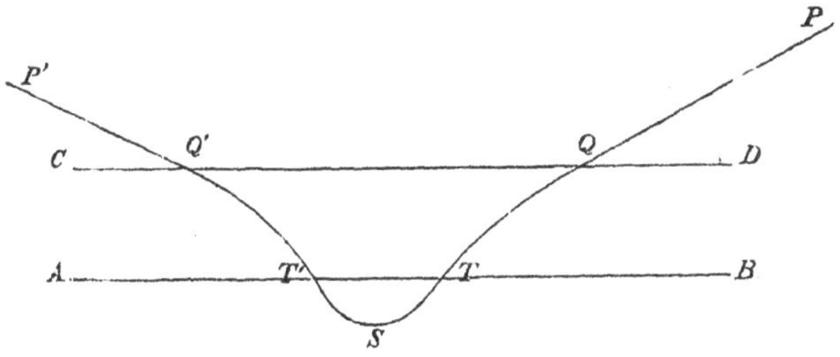

AB may be destroyed; then as before, the particle will pursue the path $ST'Q'P'$ similar and equal to $STQP$ and will be reflected at an angle equal to that of incidence and with its original velocity.

II. The particle may pass into the lower medium so far as to be independent of the action of the upper medium. In this case, it will move in a straight line as before, and the change of the square of its velocity will be independent of the path pursued. Hence if V be the velocity and α the angle, of incidence, V', α' those of refraction, we have

$$V \sin \alpha = V' \sin \alpha',$$

$$V^2 = V'^2 \pm a^2,$$

where a is a constant depending on the nature of the two media.

$$\text{Hence, } \frac{\sin \alpha}{\sin \alpha'} = \frac{V'}{V} = \sqrt{\left(1 \mp \frac{a^2}{V^2}\right)},$$

and therefore for particles of light which have the same velocity the ratio of the sines of the angles of incidence and refraction is constant.

We have introduced this example, although belonging to a theory almost exploded, as it forms a good illustration of the application of the results of this Chapter.

EXAMPLES.

(1) The time of describing any portion PQ of the parabolic path of a particle acted on by gravity, is proportional to the difference of the tangents of the angles which the tangents at P and Q make with the horizon. (Art. 71).

(2) The sights of a gun are set so that the ball may strike a given object; shew that when the sights are directed to any other object in the same vertical line, the ball will also strike it.

(3) Shew that the time of a projectile's describing an arc of its path cut off by a focal chord is equal to the time of falling vertically from rest through a space equal to the chord.

(4) If a shell burst, all the fragments receiving equal velocities from the explosion; shew that the locus of the foci of the paths of the fragments is a sphere, of the vertices an oblate spheroid, and of the particles themselves at any time a sphere.

(5) Two bodies A, B acted on by gravity are projected from two given points in the same vertical line with the same velocity and in parallel directions; shew that if A be higher than B, pairs of tangents drawn to B's path from any points of A's path will intercept arcs described by B in equal times.

(6) If v, v', v'', be the velocities at three points P, Q, R, of the path of a projectile where the inclinations to the horizon are α, $\alpha-\beta$, $\alpha-2\beta$; and if t, t' be the times of describing PQ, QR respectively, shew that

$$v''t=vt', \text{ and } \frac{1}{v}+\frac{1}{v''}=\frac{2\,.\cos\beta}{v'} \text{ (Art. 71).}$$

(7) If two particles be projected from the same point at the same instant in the same vertical plane, with velocities v and v_1 in directions making angles α and α_1 with the horizon; shew that the interval between their transits through the other point which is common to their paths is

$$\frac{2}{g}\cdot\frac{vv_1\sin(\alpha\sim\alpha_1)}{v_1\cos\alpha_1+v\cos\alpha}.$$

(8) If any chord be drawn to the trajectory of a projectile, the velocities of the particle at its extremities if resolved perpendicular to the chord, are equal. (Art. 71).

(9) Particles slide from rest at the highest point of a vertical circle down chords, and are then allowed to move freely; shew that the locus of the foci of their paths is a circle of half the dimensions, and that all the paths bisect the vertical radius.

(10) If the particles slide down chords to the lowest point, and be then suffered to move freely, the locus of the foci is a cardioide.

(11) Down what chord from the vertex of a vertical circle must a particle slide so as to have when falling freely the greatest range on a given horizontal plane?

(12) Find the locus of the foci of all trajectories which pass through two given points.

(13) The envelope of all the parabolas which correspond to a given velocity of projection is equal to the trajectory for which the direction of projection is horizontal.

(14) Particles fall down *diameters* of a vertical circle; the locus of the foci of their subsequent paths is the circle.

(15) If two bodies be projected from the same point, with equal velocities, and in such directions that they both arrive at the same point of a plane whose inclination to the horizon is β, and if t, t' be the times of flight, and α the angle of projection of the first

$$t' = t \cdot \frac{\cos\alpha}{\sin(\alpha-\beta)}.$$

(16) If the focus of the projectile's path be as much below the horizontal plane through the point of projection, as the vertex is above; shew that double the angle of projection

$$= \sec^{-1} 3.$$

(17) From points of an inclined plane, particles are simultaneously projected in different directions. If their times of flight are the same, shew that their locus at any instant is a plane parallel to the given one.

(18) A particle is thrown over a triangle from one end of the horizontal base, and, grazing the vertex, falls upon the other end of the base. If α, β be the base angles, θ the angle of projection,

$$\tan\theta = \tan\alpha + \tan\beta.$$

(19) For the greatest range on an inclined plane through the point of projection the direction of motion on leaving, is at right angles to that on reaching, the plane.

(20) Particles are projected from the same point: 1st. with the same vertical, 2nd. with the same horizontal, velocity; shew that in each case the locus of the foci is a parabola whose focus is at point of projection, and axis vertical, but whose vertex is upwards in case (1) and downwards in (2).

(21) If α be the angle of projection, T the time which elapses before the projectile strikes the ground, prove that at the time $\frac{T}{4 \sin^2 \alpha}$ the angle which the direction of motion makes with the direction of projection is $\frac{\pi}{2} - \alpha$.

(22) If a body describe an arc of a cycloid under the action of a force parallel to the base, shew that this force varies inversely as $2 \sin \theta - \sin 2\theta$, θ being the corresponding arc of the generating circle measured from the vertex.

(23) If the force perpendicular to a plane vary as the distance, shew that the curves described have equations of the form

$$\left.\begin{array}{r} y = Aa^x + Ba^{-x}, \\ \text{or } y = A \cos(mx + B) \end{array}\right\} \begin{array}{l} \text{as the force is repulsive} \\ \text{or attractive.} \end{array}$$

Find the circumstances of projection in the two cases that the curves may be the catenary, and the companion to the cycloid, respectively.

CHAPTER V.

CENTRAL FORCES.

81. In this part of the subject we consider the motion of a particle under the action of a force whose direction always passes through, and whose intensity is some function of the distance from, a fixed point. The fixed point is called the center of force, and the force is said to be attractive or repulsive according as it is directed to or from the center. The former as including the most important applications of the subject, we will take as our standard case, but it will be seen that a simple change of sign will adapt our general formulæ to the latter.

82. Consider then a particle projected in the plane of the paper, and acted on by an accelerating force P directed to the fixed point O. The whole motion will clearly take place in

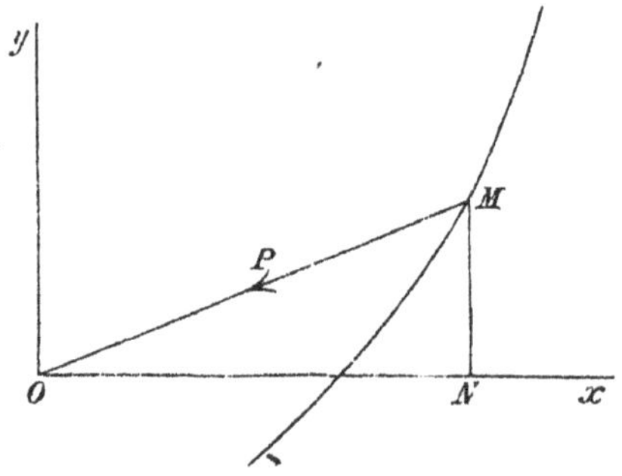

this plane as there is no force to withdraw the particle from it. Let Ox, Oy any two lines through O at right angles to each other be taken as the axes of co-ordinates. Let M be the position of the particle at the time t; and draw MN perpendicular to Ox, and join MO. Let $ON = x$, $NM = y$, $OP = r$, and the angle $NOM = \theta$. Then since $\cos\theta = \frac{x}{r}$, $\sin\theta = \frac{y}{r}$ the components of P parallel to the axes, and in the negative directions are $P\frac{x}{r}$, $P\frac{y}{r}$. But by the second law of motion we may consider the accelerations in the directions of x and y separately, and we have therefore

$$\left.\begin{aligned}\frac{d^2x}{dt^2} &= -P\frac{x}{r}\\ \frac{d^2y}{dt^2} &= -P\frac{y}{r}\end{aligned}\right\}\text{.................... } (A).$$

In these since P is a function of r, and therefore of x and y, the second members will generally contain both these variables, and the equations must be treated as simultaneous differential equations. Their integrals will give x, y, $\frac{dx}{dt}$, $\frac{dy}{dt}$ in terms of t, from which the position and velocity of the particle at any instant will be known, and the problem completely solved. In one case however when P is proportional to r, the first equation will involve x and t, and the second y and t only, and each equation may be integrated by itself. As being the simplest example of the sort we are considering, and of great importance in its applications, especially to Physical Optics, we will shortly consider it.

83. Prop. *A particle moves about a center of force, the force varying directly as the distance: to determine the motion.*

Let μ be the accelerating force at an unit of distance, usually called the absolute force of the center, then $P = \mu r$, and equations (A) become

$$\left.\begin{aligned}\frac{d^2x}{dt^2} &= -\mu x\\ \frac{d^2y}{dt^2} &= -\mu y\end{aligned}\right\}\text{.................... } (B),$$

the integrals of which, see Art.(42), are

$$x = A \cos\{\sqrt{(\mu)}\,t + B\} \;\ldots\ldots\ldots\ldots\; (1),$$

$$y = A' \cos\{\sqrt{(\mu)}\,t + B'\} \;\ldots\ldots\ldots\ldots\; (2),$$

A, B, A', B' being the constants introduced in the integration, to be determined by the initial circumstances of motion. Consider the particle projected from a point on the axis of x at distance a from the center with velocity V, and in a direction making an angle α with Ox. Then we have when $t = 0$, $x = a$, $y = 0$, $\frac{dx}{dt} = V\cos\alpha$, $\frac{dy}{dt} = V\sin\alpha$. Hence,

$$a = A \cos B$$

$$0 = A' \cos B',$$

$$V\cos\alpha = -A\sqrt{(\mu)}\sin B,$$

$$V\sin\alpha = -A'\sqrt{(\mu)}\sin B'.$$

Expanding the cosines in (1) and (2), and substituting from these equations, we obtain

$$x = \frac{V\cos\alpha}{\sqrt{(\mu)}}\sin\sqrt{(\mu)}\,.\,t + a\cos\sqrt{(\mu)}\,.\,t \;\ldots.\; (3),$$

$$y = \frac{V\sin\alpha}{\sqrt{(\mu)}}\sin\sqrt{(\mu)}\,.\,t \;\ldots\ldots\ldots\ldots\ldots\ldots\ldots\; (4),$$

which contain the complete solution of the problem. Eliminating t, we have

$$(x\sin\alpha - y\cos\alpha)^2 + \frac{\mu a^2}{V^2}y^2 = a^2\sin^2\alpha,$$

the equation to the path of the particle which is therefore an ellipse about the center O. Equations (3) and (4), give periodic values for x, y, $\frac{dx}{dt}$, $\frac{dy}{dt}$ such that all the circumstances of motion will be the same at the time $t + \frac{2\pi}{\sqrt{(\mu)}}$ as at the time t.

The period of revolution is therefore $\frac{2\pi}{\sqrt{(\mu)}}$ which is a remarkable result, as it shews that the periodic time is independent of the dimensions of the ellipse, and depends solely on the intensity of the force.

By taking μ negative in equations (B), we may apply them to the case of a repulsive force varying as the distance from O. In the integration for this supposition the sines and cosines would be replaced by exponentials, and the curve described would be an hyperbola having O as center, but the motion would not be one of revolution as the particle would necessarily always remain on the same branch of the hyperbola.

84. Recurring to equations (A), it will in all cases but the one we have just considered be more convenient to transform them to polar co-ordinates, especially as the general differential polar equation to the orbit described by a particle under the action of a central force can be determined.

85. Prop. *A particle being acted on by a central force; required to determine the polar equation to the path.*

The equations of motion are

$$\frac{d^2x}{dt^2} = -P\frac{x}{r} \quad \text{......................} \quad (1),$$

$$\frac{d^2y}{dt^2} = -P\frac{y}{r} \quad \text{......................} \quad (2).$$

Multiplying (2) by (x), and (1) by (y), and subtracting

$$x\frac{d^2y}{dt^2} - y\frac{d^2x}{dt^2} = 0.$$

Integrating

$$x\frac{dy}{dt} - y\frac{dx}{dt} = \text{constant} = h \text{ suppose.}$$

Changing the variables from x, y, to r, θ, where $x = r\cos\theta$, $y = r\sin\theta$, we get

$$r^2\frac{d\theta}{dt} = h \quad\text{.......................}\quad (3),$$

or substituting $\frac{1}{u}$ for r,

$$\frac{d\theta}{dt} = hu^2 \quad\text{.......................}\quad (4).$$

Again, $\quad x = r\cos\theta = \frac{\cos\theta}{u}$;

$$\therefore \frac{dx}{dt} = -\frac{u\sin\theta + \cos\theta\frac{du}{d\theta}}{u^2}\cdot\frac{d\theta}{dt}$$

$$= -h\left(u\sin\theta + \cos\theta\frac{du}{d\theta}\right) \text{ by (4)};$$

$$\therefore \frac{d^2x}{dt^2} = -h\left(u\cos\theta + \cos\theta\frac{d^2u}{d\theta^2}\right)\frac{d\theta}{dt}$$

$$= -h^2u^2\left(u\cos\theta + \cos\theta\frac{d^2u}{d\theta^2}\right) \text{ by (4)}.$$

But by (1),

$$\frac{d^2x}{dt^2} = -P\cos\theta.$$

Equating these values of $\frac{d^2x}{dt^2}$ and dividing by $\cos\theta$,

$$P = h^2u^2\left(\frac{d^2u}{d\theta} + u\right) \quad\text{...............}\quad (5),$$

$$\text{or } \frac{d^2u}{d\theta^2} + u - \frac{P}{h^2u^2} = 0 \quad\text{...............}\quad (6).$$

This is the differential equation to the orbit described, and as in any particular instance P will be given in terms of r, its integral will be the polar equation to the required path.

86. The general integrals of equations (1) and (2) which are differential equations of the second order, ought to contain four constants. One of these has been already introduced in (3), and two more will be introduced by the integration of (5). If the value of r deduced from the integral of (5) be substituted in (3) and the equation then integrated, the remaining constant will be introduced, and both the path of the particle and its position at any time will be obtained. The four constants involved in the resulting equations must be determined by the initial circumstances of motion, namely, the initial position of the particle, its velocity and direction of projection.

87. Equation (5) may be used to ascertain the law of force which must act upon a particle to cause it to describe a given curve. To effect this we must determine the relation between u and θ from the equation to the orbit: we must then differentiate u twice with respect to θ, and substitute the result in the expression for P; eliminating θ, if it occur, by means of the relation between u and θ. In this way we shall obtain P in terms of u alone, and therefore of r alone.

When we know the relation between r and θ from (6), we make use of equation (3) to determine the time of describing a given portion of the orbit, or conversely to find the position of the body in its orbit at any time.

88. The equation to the orbit between r and p, the radius vector and the perpendicular on the tangent at any point, may be easily obtained from (6). For (TODHUNTER, *Diff. Calc.* Art. 295) we have

$$\frac{d^2u}{d\theta^2} + u = \frac{1}{p^3u^2} \cdot \frac{dp}{dr};$$

$$\therefore\ P = \frac{h^2}{p^3} \cdot \frac{dp}{dr}.$$

89. PROP. *The sectorial area swept out by the radius vector of the particle in any time is proportional to the time.*

If A denote this area we have, (TODHUNTER, *Diff. Calc.* Art. 313),

$$\frac{dA}{dt} = \frac{r^2}{2}\frac{d\theta}{dt};$$

and therefore by equation (3) in Art. (85),

$$\frac{dA}{dt} = \frac{h}{2},$$

$$A = \tfrac{1}{2}ht + C = \tfrac{1}{2}ht,$$

C being zero as A and t are supposed to vanish together. Let A' be the area described in any other time t', then

$$A' = \tfrac{1}{2}ht';$$

$$\therefore A : A' :: t : t';$$

or the areas described in different times are proportional to the times of describing them. We also see by taking $t = 1$, that $h =$ twice the area described in a unit of time.

90. PROP. *The velocity of the particle at different points of its path is inversely proportional to the perpendiculars from the center of force on the tangents at those points.*

For $$\text{velocity} = v = \frac{ds}{dt}$$

$$= \frac{ds}{d\theta}\cdot\frac{d\theta}{dt}$$

$$= \frac{r^2}{p}\cdot\frac{d\theta}{dt}, \text{ by } \textit{Diff. Calc.}$$

(where p is the perpendicular on the tangent)

$$=\frac{h}{p}, \text{ by equation (3) Art. (85),}$$

$$\text{or } v \propto \frac{1}{p}.$$

91. This equation enables us to express h in terms of the initial circumstances of motion. For let R be the distance of the point of projection from the center, V the velocity, and β the angle which the direction of projection makes with R. Then evidently

perpendicular on tangent at point of projection $= R \sin\beta$;

$$\therefore V = \frac{h}{R\sin\beta},$$

$$\text{or } h = VR\sin\beta.$$

Again, since (*Diff. Calc.* Art. 284)

$$\frac{1}{p^2} = u^2 + \left(\frac{du}{d\theta}\right)^2,$$

we have

$$v^2 = \frac{h^2}{p^2} = h^2\left\{u^2 + \left(\frac{du}{d\theta}\right)^2\right\},$$

another important expression for the velocity.

92. PROP. *The velocity at any point of a central orbit is independent of the path described, and depends solely on the intensity and law of the force, the distance of the point from the center, and the velocity and distance of projection.*

Multiply equations (A) by $2\frac{dx}{dt}$, $2\frac{dy}{dt}$ respectively, and add, then

$$2\frac{dx}{dt}\cdot\frac{d^2x}{dt^2} + 2\frac{dy}{dt}\cdot\frac{d^2y}{dt^2} = -2\frac{P}{r}\left(x\frac{dx}{dt} + y\frac{dy}{dt}\right)$$

$$= -2P\frac{dr}{dt}.$$

Since $x^2 + y^2 = r^2$, and $\therefore\ x\frac{dx}{dt} + y\frac{dy}{dt} = r\frac{dr}{dt}$.

But $v^2 = \left(\frac{ds}{dt}\right)^2 = \left(\frac{dx}{dt}\right)^2 + \left(\frac{dy}{dt}\right)^2$;

$$\therefore\ \frac{d(v^2)}{dt} = -2P\frac{dr}{dt}.$$

Since P is some function of r, let $P = \phi(r)$, then

$$v^2 = C - 2\int \phi(r)\, dr$$

$$= C - 2f(r), \text{ suppose.}$$

At the point of projection $v = V$, $r = R$;

$$\therefore\ V^2 = C - 2f(R);$$

$$\therefore\ v^2 - V^2 = 2\{f(R) - f(r)\},$$

which proves the proposition.

93. PROP. *The velocity of a particle at any point of a central orbit is the same as would be acquired by a particle moving freely from rest along one-fourth of the chord of curvature at the point drawn through the center of force, under the action of a constant force whose intensity is equal to that of the central force at the point.*

By last Art.,

$$\frac{d(v^2)}{dt} = -2P\frac{dr}{dt};$$

$$\therefore\ v\frac{dv}{dr} = -P.$$

And by Art. (90),

$$v = \frac{h}{p}.$$

Differentiating the logarithm of the latter

$$\frac{1}{v}\cdot\frac{dv}{dr} = -\frac{1}{p}\cdot\frac{dp}{dr},$$

and dividing the former equation by this

$$v^2 = Pp\frac{dr}{dp} = 2P.\frac{p}{r}.\frac{2r}{4}\frac{dr}{dp}$$

$$= 2P.\frac{q}{4}, \text{ (Diff. Calc. 321)},$$

where q is the chord of curvature through the center. Hence the proposition, Art. (36).

From this equation the velocity V of a particle moving in a circle of radius R under the action of any force P to the center is given by the equation

$$V^2 = P.R.$$

94. Def. An apse is a point in a central orbit at which the tangent is at right angles to the radius vector, and the radius at such a point is called an apsidal distance.

The analytical condition for such a point (*Diff. Calc.* 279), is $\frac{du}{d\theta} = 0$, and shews that an apsidal distance is a maximum or minimum value of the radius vector.

Prop. *Every apsidal line divides the orbit into two parts which are equal and similar.*

For $\frac{du}{d\theta}$ changes sign in passing through an apse, and therefore since $\left(\frac{du}{d\theta}\right)^2$ is some function of u, $= f(u)$ suppose, if on one side of the apse $\frac{du}{d\theta}$ be represented by $+\sqrt{\{f(u)\}}$, on the other it will be represented by $-\sqrt{\{f(u)\}}$.

Let then A be an apse, O the center of force, and OP, OQ

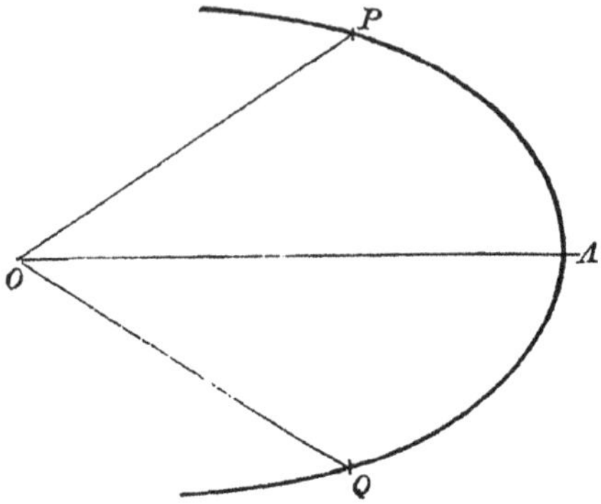

any two lines on opposite sides of OA and equally inclined to it, then

$$\angle\, QOA = \int_{\frac{1}{OQ}}^{\frac{1}{OA}} \frac{du}{\sqrt{\{f(u)\}}},$$

$$\angle\, AOP = -\int_{\frac{1}{OA}}^{\frac{1}{OP}} \frac{du}{\sqrt{\{f(u)\}}};$$

$$\therefore \int_{\frac{1}{OQ}}^{\frac{1}{OA}} \frac{du}{\sqrt{\{f(u)\}}} + \int_{\frac{1}{OA}}^{\frac{1}{OP}} \frac{du}{\sqrt{\{f(u)\}}} = 0,$$

$$\text{or } \int_{\frac{1}{OQ}}^{\frac{1}{OP}} \frac{du}{\sqrt{\{f(u)\}}} = 0 \text{ identically};$$

$$\therefore OP = OQ,$$

which shews that any two lines on opposite sides of and equally inclined to OA are equal. Hence the parts into which OA divides the orbit are equal and similar.

95. PROP. *In a central orbit there cannot be more than two apsidal distances.*

For since the parts of the orbit on opposite sides of an apse are similar, the particle after passing two apses must come next to one at an equal distance to that of the first, then to one at an equal distance to that of the second, and so on. Hence there can only be two apsidal distances.

96. When the central force varies as any power of the distance, we may obtain the same result from equation (6), Art. (85), as well as the equation for determining the apsidal distances. Suppose $P = \mu u^n$, then we have

$$\frac{d^2u}{d\theta^2} + u - \frac{\mu}{h^2} u^{n-2} = 0.$$

Multiplying by $2h^2 \frac{du}{d\theta}$ and integrating

$$h^2 \left\{ \left(\frac{du}{d\theta}\right)^2 + u^2 \right\} = v^2 = \frac{2\mu}{n-1} u^{n-1} + C.$$

Suppose the particle projected with a velocity equal to q times the velocity from infinity at the same distance, and let c be the initial value of u, then when $u = c$,

$$v^2 = \frac{2\mu q^2}{n-1} c^{n-1} \text{ (Art. 55)};$$

$$\therefore C = (q^2 - 1) \frac{2\mu}{n-1} c^{n-1};$$

$$\therefore h^2 \left\{ \left(\frac{du}{d\theta}\right)^2 + u^2 \right\} = \frac{2\mu}{n-1} \{u^{n-1} + (q^2 - 1) c^{n-1}\}.$$

To determine the apsidal distances we must put $\frac{du}{d\theta} = 0$, which gives

$$u^{n-1} - \frac{h^2 (n-1)}{2\mu} u^2 + (q^2 - 1) c^{n-1} = 0.$$

The form of this equation shews that it can have (at

most) two positive roots, which are therefore the two apsidal distances.

Although there can only be two apsidal distances there may be any number of apses, and the angle between two consecutive apsidal distances is called the apsidal angle. Generally to determine this angle the equation to the orbit must first be found for the particular case considered, but it may be determined approximately when the orbit does not differ much from a circle.

97. Prop. *A particle revolves in an orbit which is very nearly circular, acted on by a central force varying as any function of the distance, to determine the apsidal angle.*

Assume $P = \mu u^2 \phi(u)$, then

$$\frac{d^2u}{d\theta^2} + u - \frac{\mu}{h^2}\phi(u) = 0.$$

But since the orbit is very nearly circular, u may be put $u = c + x$, where c is constant and x always small. Hence

$$\frac{d^2x}{d\theta^2} + c + x - \frac{\mu}{h^2}\phi(c+x) = 0,$$

$$\text{or } \frac{d^2x}{d\theta^2} + c + x - \frac{\mu}{h^2}\{\phi(c) + x\phi'(c)\} = 0 \text{ nearly.}$$

But Art. (91) $h^2 = V^2 R^2 \sin^2\beta$, and as the orbit is nearly a circle radius $\frac{1}{c}$ we will have

$$h^2 = \frac{V^2}{c^2} \text{ nearly,}$$

where V is the velocity in a circle at the distance $\frac{1}{c}$, the central force being $\mu c^2 \phi(c)$;

$$\therefore\ V^2 = \frac{\mu c^2 \phi(c)}{c}, \text{ Art. (93);}$$

$$\therefore h^2 = \frac{\mu\phi(c)}{c}, \text{ or } \frac{\mu}{h^2} = \frac{c}{\phi(c)}.$$

Substituting,

$$\frac{d^2x}{d\theta^2} + \left\{1 - \frac{c\phi'(c)}{\phi(c)}\right\} x = 0,$$

the integral of which is, Art. (42),

$$x = A \cos\left[\sqrt{\left\{1 - \frac{c\phi'(c)}{\phi(c)}\right\}} . \theta + B\right].$$

Hence the general value of θ which renders $\frac{dx}{d\theta} = 0$, is given by the equation

$$\sqrt{\left\{1 - \frac{c\phi'(c)}{\phi(c)}\right\}} . \theta + B = n\pi,$$

n being any integer, and consequently the difference between any two successive values of θ

$$= \frac{\pi}{\sqrt{\left\{1 - \frac{c\phi'(c)}{\phi(c)}\right\}}},$$

the apsidal angle approximately.

Thus if the force varied directly as the nth power of the distance, we should have

$$\mu u^2 \phi(u) = \mu u^{-n}; \quad \therefore \phi(u) = u^{-n-2},$$

$$\phi'(u) = -(n+2) u^{-n-3},$$

$$\text{and } \therefore \text{ apsidal angle} = \frac{\pi}{\sqrt{(3+n)}}.$$

This shews that n cannot be less than -3, or that the force cannot vary according to a higher inverse power of the distance than the third, in order that the circle may be an approximation to the path of the particle.

If we wish to find the law of force that the apsidal angle may be given $=\alpha$ suppose, we have

$$\frac{\pi}{\sqrt{\left\{1-\frac{c\phi'(c)}{\phi(c)}\right\}}}=\alpha;$$

$$\therefore \frac{\phi'(c)}{\phi(c)}=\frac{1}{c}\left(1-\frac{\pi^2}{\alpha^2}\right);$$

$$\therefore \log\frac{\phi(c)}{C}=\left(1-\frac{\pi^2}{\alpha^2}\right)\log c,$$

$$\text{or } \phi(c)=Cc^{1-\frac{\pi^2}{\alpha^2}};$$

therefore the law of force is $u^2\phi(u)$ or $\mu u^{3-\frac{\pi^2}{\alpha^2}}$.

98. PROP. *A particle is projected from a given point in a given direction and with a given velocity, and moves under the action of a central force varying inversely as the square of the distance; to determine the orbit.*

We have $P=\mu u^2$, and therefore

$$\frac{d^2u}{d\theta^2}+u-\frac{\mu}{h^2}=0,$$

$$\text{or } \frac{d^2}{d\theta^2}\left(u-\frac{\mu}{h^2}\right)+u-\frac{\mu}{h^2}=0;$$

the integral of which is

$$u-\frac{\mu}{h^2}=A\cos(\theta+B),$$

or as it is usually written

$$u=\frac{\mu}{h^2}\{1+e\cos(\theta-\alpha)\}\ldots\ldots\ldots\ldots(1);$$

$$\therefore \frac{du}{d\theta}=-\frac{\mu}{h^2}e\sin(\theta-\alpha).$$

Let R be the distance of the point of projection from the center, β the angle, and V the velocity of projection; then when $\theta = 0$,

$$u = \frac{1}{R}, \quad \cot\beta = -\left(\frac{1}{u}\cdot\frac{du}{d\theta}\right)_{\theta=0} \quad (\textit{Diff. Calc.}\ 279);$$

$$\therefore \frac{h^2}{\mu R} - 1 = e\cos\alpha,$$

$$\frac{h^2}{\mu R}\cot\beta = -e\sin\alpha;$$

$$\therefore \tan\alpha = \frac{h^2\cot\beta}{\mu R - h^2},$$

$$e^2 = \frac{h^4}{\mu^2 R^2}\operatorname{cosec}^2\beta - \frac{2h^2}{\mu R} + 1.$$

But $h^2 = V^2 R^2 \sin^2\beta$, Art. (91);

$$\therefore \tan\alpha = \frac{V^2 R\sin\beta\cos\beta}{\mu - V^2 R\sin^2\beta} \text{ .... } (2),$$

$$1 - e^2 = \frac{V^2 R^2 \sin^2\beta}{\mu}\left(\frac{2}{R} - \frac{V^2}{\mu}\right) \text{ ....} (3).$$

Now (1) is the general equation to a conic section about the focus, and its nature will depend on the value of e given by (3), we see therefore that

if $V^2 > \dfrac{2\mu}{R}$, $e > 1$, and the orbit is an hyperbola,

$V^2 = \dfrac{2\mu}{R}$, $e = 1$, ... a parabola,

$V^2 < \dfrac{2\mu}{R}$, $e < 1$, ... an ellipse.

99. By Art. (55) the square of the velocity from infinity at distance R for the law of force we are considering is $\dfrac{2\mu}{R}$,

and the above conditions may therefore be expressed more concisely by saying that the orbit will be an hyperbola, a parabola, or an ellipse, according as the velocity of projection is greater than, equal to, or less than the velocity from infinity.

The velocity of a particle moving in a circle is also often taken as the standard of comparison for estimating the velocities of bodies in their orbits. For the present law of force the square of the velocity in a circle at distance R is $\frac{\mu}{R}$, and the above conditions may also be expressed by saying that the orbit will be an hyperbola, a parabola, or an ellipse, according as the velocity of projection is greater than, equal to, or less than $\sqrt{(2)}$ times the velocity in a circle at the same distance.

100. Supposing the orbit an ellipse, we shall obtain the axis major and latus rectum of the orbit best by a different process of integrating the equation. Multiplying it by $2h^2\frac{du}{d\theta}$ and integrating, we obtain

$$h^2\left\{\left(\frac{du}{d\theta}\right)^2 + u^2\right\} = v^2 = C + 2\mu u.$$

But when $u = \frac{1}{R}$, $v = V$; $\therefore\ C = V^2 - \frac{2\mu}{R}$;

$$\therefore\ h^2\left\{\left(\frac{du}{d\theta}\right)^2 + u^2\right\} = v^2 = V^2 - \frac{2\mu}{R} + 2\mu u \ldots\ldots\ldots (4).$$

Now to determine the apsidal distances, we must put

$$\frac{du}{d\theta} = 0;$$

$$\therefore\ u^2 - \frac{2\mu}{h^2}u + \frac{2\mu}{h^2 R} - \frac{V^2}{h^2} = 0,$$

is a quadratic equation the roots of which are the reciprocals of the two apsidal distances. But if a be the semiaxis major, and e the eccentricity, these distances are

$a(1-e)$ and $a(1+e)$: Hence

$$\frac{1}{a(1-e)}+\frac{1}{a(1+e)}=\frac{2\mu}{h^2};$$

$$\text{or } a(1-e^2)=\frac{h^2}{\mu} \quad\ldots\ldots\ldots\ldots\ldots\ldots (5).$$

And

$$\frac{1}{a^2(1-e^2)}=\frac{2\mu}{h^2R}-\frac{V^2}{h^2};$$

$$\therefore \frac{1}{a}=\frac{2}{R}-\frac{V^2}{\mu} \quad\ldots\ldots\ldots\ldots (6).$$

Substituting then $\frac{\mu}{a}$ for $\frac{2\mu}{R} - V^2$ in (4), we have

$$v^2=\mu\left(2u-\frac{1}{a}\right) \quad\ldots\ldots\ldots\ldots\ldots\ldots (7).$$

Equations (5) and (6) give the latus rectum and axis major of the orbit, and shew that the axis major is independent of the direction of projection.

Equation (7) gives a useful expression for the velocity at any point.

101. The time of describing any given angle is to be obtained from the formula,

$$r^2\frac{d\theta}{dt}=h$$

$$=\sqrt{\{\mu a(1-e^2)\}} \text{ by equation (5)};$$

$$\therefore \frac{dt}{d\theta}=\frac{r^2}{\sqrt{\{\mu a(1-e^2)\}}}$$

$$=\sqrt{\left(\frac{a^3(1-e^2)^3}{\mu}\right)}\frac{1}{(1+e\cos\theta)^2};$$

measuring the angle from the nearer apse. To integrate this let

$$\Theta = \frac{\sin\theta}{1 + e\cos\theta}, \text{ then}$$

$$\frac{d\Theta}{d\theta} = \frac{\cos\theta + e}{(1 + e\cos\theta)^2} = \frac{\frac{1}{e}(1 + e\cos\theta) - \frac{1-e^2}{e}}{(1 + e\cos\theta)^2}$$

$$= \frac{1}{e}\cdot\frac{1}{1 + e\cos\theta} - \frac{1-e^2}{e}\,\frac{1}{(1 + e\cos\theta)^2};$$

$$\therefore \int\frac{d\theta}{(1+e\cos\theta)^2} = -\frac{e\Theta}{1-e^2} + \frac{1}{1-e^2}\int\frac{d\theta}{1 + e\cos\theta}$$

$$= -\frac{e}{1-e^2}\cdot\frac{\sin\theta}{1+e\cos\theta} + \frac{1}{1-e^2}\int\frac{\sec^2\frac{\theta}{2}\,d\theta}{(1+e)+(1-e)\tan^2\frac{\theta}{2}};$$

$$= -\frac{e}{1-e^2}\,\frac{\sin\theta}{1 + e\cos\theta} + \frac{2}{(1-e^2)^{\frac{3}{2}}}\tan^{-1}\left\{\sqrt{\left(\frac{1-e}{1+e}\right)}\cdot\tan\frac{\theta}{2}\right\},$$

if $e < 1$;

$$\text{or } = \frac{e}{e^2-1}\cdot\frac{\sin\theta}{1+e\cos\theta} - \frac{1}{(e^2-1)^{\frac{3}{2}}}\log\left\{\frac{\sqrt{(e+1)}\cos\frac{\theta}{2} + \sqrt{(e-1)}\sin\frac{\theta}{2}}{\sqrt{(e+1)}\cos\frac{\theta}{2} - \sqrt{(e-1)}\sin\frac{\theta}{2}}\right\}$$

if $e > 1$.

Hence the time of describing about the focus an angle θ measured from the nearer apse is

$$\sqrt{\frac{a^3}{\mu}}\left[2\tan^{-1}\left\{\sqrt{\left(\frac{1-e}{1+e}\right)}\tan\frac{\theta}{2}\right\} - e\sqrt{(1-e^2)}\,\frac{\sin\theta}{1 + e\cos\theta}\right],$$

in the ellipse, and

$$\sqrt{\frac{a^3}{\mu}}\left[\log\left\{\frac{\sqrt{(e+1)}\cos\frac{\theta}{2}-\sqrt{(e-1)}\sin\frac{\theta}{2}}{\sqrt{(e+1)}\cos\frac{\theta}{2}+\sqrt{(e-1)}\sin\frac{\theta}{2}}\right\}+e\sqrt{(e^2-1)}\frac{\sin\theta}{1+e\cos\theta}\right],$$

in the hyperbola.

102. In the parabola if d be the apsidal distance the integral becomes

$$\{\text{since } e=1,\quad a(1-e)=d,\quad a(1-e^2)=2d\},$$

$$t=\sqrt{\left(\frac{8d^3}{\mu}\right)}\int\frac{d\theta}{(1+\cos\theta)^2}$$

$$=\sqrt{\left(\frac{8d^3}{\mu}\right)}\int\frac{1}{4}\sec^4\frac{\theta}{2}\,d\theta$$

$$=\sqrt{\left(\frac{2d^3}{\mu}\right)}\int\left(1+\tan^2\frac{\theta}{2}\right)d\tan\frac{\theta}{2}$$

$$=\sqrt{\left(\frac{2d^3}{\mu}\right)}\left(\tan\frac{\theta}{2}+\frac{1}{3}\tan^3\frac{\theta}{2}\right).$$

103. From the result for the ellipse we see that the periodic time $=2\pi\sqrt{\frac{a^3}{\mu}}$, which may also be found from the consideration that the periodic time

$$=\frac{2\text{ area of ellipse}}{h}=\frac{2\pi a^2\sqrt{(1-e^2)}}{\sqrt{\{\mu a(1-e^2)\}}}$$

$$=2\pi\sqrt{\frac{a^3}{\mu}}.$$

104. By laborious calculation from an immense series of observations of the planets and of Mars in particular, Kepler enunciated the following, as the laws of the planetary motions about the Sun.

I. The planets describe Ellipses of which the Sun occupies a focus.

II. The radius vector of each planet traces out equal areas in equal times.

III. The squares of the periodic times of any two planets, are as the cubes of the major axes of their orbits.

105. From the second of these laws we conclude that the planets are retained in their orbits by a central force tending to the Sun. For,

Prop. *If the radius vector of a particle moving in a plane describe equal areas in equal times about a point in that plane, the resultant force on the particle tends to that point.*

Take the point as origin, and let x, y be the co-ordinates of the particle at time t, X, Y the forces acting on it resolved parallel to the axes; the equations of motion are

$$\frac{d^2x}{dt^2} = X, \quad \frac{d^2y}{dt^2} = Y \ldots\ldots\ldots\ldots\ldots\ldots (1).$$

But by hypothesis, if A be the area traced out by the radius vector, $\frac{dA}{dt}$ is constant.

Hence, $$2\frac{dA}{dt} = x \cdot \frac{dy}{dt} - y \cdot \frac{dx}{dt} = C.$$

Differentiating, $$x \cdot \frac{d^2y}{dt^2} - y \cdot \frac{d^2x}{dt^2} = 0;$$

or by (1) $$xY - yX = 0.$$

Hence, $\frac{Y}{X} = \frac{y}{x}$, and by the parallelogram of forces the resultant of X and Y passes through the origin.

106. From the first it follows that the law of the force is that of the inverse square of the distance.

The equation to an Ellipse about the focus is

$$u = \frac{2}{l}(1 + e\cos\theta), \text{ where } l \text{ is the latus rectum.}$$

Hence, $\frac{d^2u}{d\theta^2} = -\frac{2e}{l}\cos\theta$, and therefore the force to the focus requisite for its description is (Art. 87)

$$P = h^2u^2\left(\frac{d^2u}{d\theta^2} + u\right)$$

$$= \frac{2h^2}{l} \cdot u^2;$$

and varies therefore inversely as the square of the distance.

107. From the third it follows that the force towards the Sun which acts on each of the planets is the same for each planet at the same distance.

Since in the formula in (103), T^2 will not vary as a^3 unless μ be constant, i.e. unless the *absolute* force of the Sun be the same for all the planets.

We shall find afterwards (Chap. XII.) that these laws are only approximate for more reasons than one, but their enunciation was sufficient to enable Newton at once to propound the doctrine of Universal Gravitation; viz. that every particle of matter in the universe attracts every other with a force which is as its mass directly, and as the square of the distance inversely.

EXAMPLES.

(1) A particle describes an ellipse under the action of a force always directed to the center, to determine the law of the force.

From the polar equation to the ellipse, center pole

$$u^2 = \frac{\cos^2\theta}{a^2} + \frac{\sin^2\theta}{b^2}; \quad \therefore\ u\frac{du}{d\theta} = \left(\frac{1}{b^2} - \frac{1}{a^2}\right)\cos\theta\sin\theta;$$

$$\therefore\ u\frac{d^2u}{d\theta^2} + \left(\frac{du}{d\theta}\right)^2 = \left(\frac{1}{b^2} - \frac{1}{a^2}\right)(\cos^2\theta - \sin^2\theta);$$

$$\therefore\ P = \frac{h^2}{u}\left(u^4 + u^3\frac{d^2u}{d\theta^2}\right)$$

$$= \frac{h^2}{u}\left[u^4 - u^2\left(\frac{du}{d\theta}\right)^2 + u^2\left\{u\frac{d^2u}{d\theta^2} + \left(\frac{du}{d\theta}\right)^2\right\}\right]$$

$$=\frac{h^2}{u}\left\{\left(\frac{\cos^2\theta}{a^2}+\frac{\sin^2\theta}{b^2}\right)^2-\left(\frac{1}{b^2}-\frac{1}{a^2}\right)^2\cos^2\theta\sin^2\theta\right.$$

$$\left.+\left(\frac{\cos^2\theta}{a^2}+\frac{\sin^2\theta}{b^2}\right)\left(\frac{1}{b^2}-\frac{1}{a^2}\right)(\cos^2\theta-\sin^2\theta)\right\}$$

$$=\frac{h^2}{u}\left(\frac{\cos^4\theta+2\cos^2\theta\sin^2\theta+\sin^4\theta}{a^2b^2}\right)=\frac{h^2}{a^2b^2}r;$$

and therefore the law is that of the direct distance.

The above example is the converse of Art. (83).

(2) A particle describes a conic section under the action of a force always directed to one of the foci, to find the law of force.

In this case

$$u=\frac{1}{a(1-e^2)}\{1+e\cos(\theta-\alpha)\},$$

$$\text{and } \therefore P=h^2u^2\left(\frac{d^2u}{d\theta^2}+u\right)$$

$$=\frac{h^2u^2}{a(1-e^2)}\propto\frac{1}{r^2},$$

the converse of Art. (98).

(3) Find the law of force tending to the pole under which a particle may describe an equiangular spiral.

Ans. $P\propto\frac{1}{r^3}$.

(4) Find the law of force by which a particle may describe a lemniscate, the center of force being the node of the lemniscate.

Ans. $P\propto\frac{1}{r^7}$.

(5) Find the law of force by which a particle may describe a circle, the center of force being in the circumference of the circle.

Ans. $P\propto\frac{1}{r^5}$.

(6) Find the law of force by which a particle may describe the spiral $r = a\left(\sec\frac{\theta}{n}\right)^n$, the center of force being the pole of the spiral.

Ans. $P \propto \dfrac{1}{r^{\frac{3n-2}{n}}}$.

(7) Find the law of force that a particle may describe the cissoid of Diocles $r = 2a\dfrac{\sin^2\theta}{\cos\theta}$, the center of force being the pole.

Ans. $P \propto \dfrac{\operatorname{cosec}^2\theta}{r^3}$.

(8) A particle is projected from a given point in a given direction with the velocity which it would acquire in falling directly towards the center to the point of projection from an infinite distance, and is acted on by a force varying inversely as the nth power of the distance, to determine the orbit.

Here $P = \mu u^n$, and therefore Art. (85)

$$\frac{d^2u}{d\theta^2} + u - \frac{\mu}{h^2}u^{n-2} = 0.$$

Multiplying by $2h^2\dfrac{du}{d\theta}$ and integrating

$$h^2\left\{\left(\frac{du}{d\theta}\right)^2 + u^2\right\} = v^2 = C + \frac{2\mu}{n-1}u^{n-1}.$$

Let V be the initial velocity, β the angle of projection, and c the initial value of u, then Art. (55)

$$V^2 = \frac{2\mu}{n-1}c^{n-1}.$$

But when $u = c$, $v = V$; $\therefore C = 0$, and

$$\therefore \left(\frac{du}{d\theta}\right)^2 + u^2 = \frac{2\mu u^{n-1}}{h^2(n-1)}.$$

Now Art. (91),

$$h^2 = \frac{V^2 \sin^2 \beta}{c^2} = \frac{2\mu}{n-1} c^{n-3} \sin^2 \beta;$$

$$\therefore \frac{\mu}{h^2} = \frac{n-1}{2c^{n-3} \sin^2 \beta};$$

$$\therefore \left(\frac{du}{d\theta}\right)^2 + u^2 = \frac{u^{n-1}}{c^{n-3} \sin^2 \beta};$$

$$\therefore \frac{d\theta}{du} = \frac{c^{\frac{n-3}{2}} \sin \beta}{u \sqrt{(u^{n-3} - c^{n-3} \sin^2 \beta)}} \quad \ldots\ldots\ldots\ldots (1).$$

To integrate this let

$$u^{n-3} = x^2;$$

$$\therefore (n-3)\, u^{n-4} \frac{du}{dx} = 2x.$$

Dividing, $$\frac{n-3}{u} \cdot \frac{du}{dx} = \frac{2}{x}.$$

Substituting in (1),

$$\frac{d\theta}{dx} = \frac{2}{n-3} \cdot \frac{c^{\frac{n-3}{2}} \sin \beta}{x \sqrt{(x^2 - c^{n-3} \sin^2 \beta)}}.$$

Integrating,

$$\theta + C' = \frac{2}{n-3} \sec^{-1} \frac{x}{c^{\frac{n-3}{2}} \sin \beta}$$

$$= \frac{2}{n-3} \sec^{-1} \frac{u^{\frac{n-3}{2}}}{c^{\frac{n-3}{2}} \sin \beta}.$$

Suppose we take the initial line so that $C' = 0$, then

$$\frac{u^{\frac{n-3}{2}}}{c^{\frac{n-3}{2}} \sin\beta} = \sec\frac{n-3}{2}.\theta;$$

or if R be the initial distance $= \frac{1}{c}$

$$\left(\frac{r}{R}\right)^{\frac{n-3}{2}} = \operatorname{cosec}\beta \cos\frac{n-3}{2}\theta,$$

the polar equation to the required path.

(9) A particle acted on by a central force varying inversely as the fifth power of the distance is projected in any direction with the velocity from infinity; find the orbit.

Ans. $r = R \operatorname{cosec}\beta \sin(\beta - \theta)$, β being the angle of projection, and the line joining the point of projection with the center initial line.

(10) A particle acted on by a central force varying inversely as the fifth power of the distance is projected from a given point with a velocity which is to the velocity from infinity :: 5 : 3, in a direction making an angle $\sin^{-1}\frac{2\sqrt{(6)}}{5}$ with the radius vector; find the orbit.

Here we have

$$\frac{d^2u}{d\theta^2} + u - \frac{\mu}{h^2}u^3 = 0;$$

$$\therefore h^2\left\{\left(\frac{du}{d\theta}\right)^2 + u^2\right\}, \text{ or } v^2 = C + \frac{\mu u^4}{2}.$$

But if V be the velocity of projection, c the initial value of u,

$$V^2 = \frac{25}{9}.\frac{\mu c^4}{2};$$

and when $u = c,\ v = V;\ \therefore\ C = \frac{8\mu c^4}{9}$;

$$\therefore\ \left(\frac{du}{d\theta}\right)^2 + u^2 = \frac{\mu}{h^2}\left(\frac{8c^4}{9} + \frac{u^4}{2}\right).$$

But
$$h^2 = \frac{V^2 \sin^2 \beta}{c^2} = \frac{25\mu c^4}{18c^2} \cdot \frac{24}{25};$$

$$\therefore\ \frac{\mu}{h^2} = \frac{3}{4c^2}.$$

Substituting and integrating we find after the necessary reductions

$$r = \frac{\sqrt{3}}{2} R \frac{1 - \epsilon^{\sqrt{2}\,(\theta + \alpha)}}{1 + \epsilon^{\sqrt{2}\,(\theta + \alpha)}};$$

where R is the initial distance and α a constant to be determined by the position of the initial line.

(11) A particle acted on by a force, varying partly as the inverse third and partly as the inverse fifth power of the distance is projected with the velocity from infinity at an angle with the distance, the tangent of which $= \sqrt{(2)}$, the forces being equal at the point of projection; determine the orbit.

Ans. $R - r = \frac{1}{\sqrt{(2)}} R\theta.$

(12) The force tending to the center of a circle whose radius is a being $\mu\left(r + \frac{2a^3}{r^2}\right)$, find the velocity with which a particle will describe the circle, and shew that if the velocity be suddenly doubled it will come to an apse at the distance $3a$.

(13) If $P = 2\mu \frac{u^5}{c^2} + \mu u^3$, and a particle be projected at an

angle of 45° with the initial distance $R=\frac{1}{c}$, with a velocity which is to the velocity in a circle at the same distance as $\sqrt{(2)} : \sqrt{(3)}$, find the curve described.

Ans. $r=R(1-\theta)$.

(14) If a particle be acted on by a central force varying inversely as the seventh power of the distance, and be projected from an apse with a velocity which is to the velocity in a circle at the same distance as $1 : \sqrt{(3)}$; find the equation to the curve described.

Ans. $r^2=R^2 \cos 2(\theta+\alpha)$.

(15) A particle acted on by a force varying inversely as the cube of the distance, is projected from a given point with any velocity in any direction, to separate the curves according to the circumstances of projection.

The equation of motion is

$$\frac{d^2u}{d\theta^2}+u-\frac{\mu}{h^2}.u=0 \text{ .... (1)}.$$

Let $\frac{\mu}{h^2}$ be >1, and let $\frac{\mu}{h^2}-1=k^2$; then

$$\frac{d^2u}{d\theta^2}-k^2u=0,$$

the integral of which is

$$u=A\epsilon^{k\theta}+B\epsilon^{-k\theta} \text{ .... (2)}.$$

This will resolve itself into three distinct species of curves according to the values of A and B.

SPECIES I. Let A, B have the same sign; then

$$u=A\epsilon^{k\theta}+B\epsilon^{-k\theta};$$

$$\therefore \frac{du}{d\theta}=k(A\epsilon^{k\theta}-B\epsilon^{-k\theta}).$$

The values of A and B may in these equations be expressed in terms of the initial distance, and angle of projection; but we may put the equation to the curve in a simpler form as follows. Let α be the value of θ corresponding to an apse, then when $\theta = \alpha$, $\frac{du}{d\theta} = 0$;

$$\therefore\ 0 = A\epsilon^{k\alpha} - B\epsilon^{-k\alpha},$$

which always gives a possible value of α;

$$\therefore\ A\epsilon^{k\alpha} = B\epsilon^{-k\alpha} = c \text{ suppose.}$$

Substituting $\quad u = c\left\{\epsilon^{k(\theta-\alpha)} + \epsilon^{-k(\theta-\alpha)}\right\}.$

Hence when $\theta = \alpha$, $u = 2c$, or $\frac{1}{2c}$ is the apsidal distance. As θ increases u increases or r diminishes, and when $\theta = \infty$, $u = \infty$, or $r = 0$, and as the curve is symmetrical on both sides of the apse it will be as represented in the figure, where A is the apse and O the center of force.

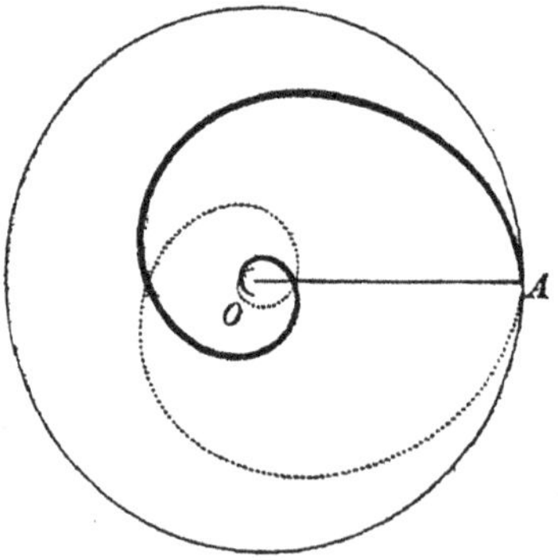

SPECIES II. Let $\frac{\mu}{h^2} > 1$, $B = 0$, then the equation (2) becomes

$$u = A\epsilon^{k\theta}$$

the equation to the logarithmic spiral. The curve will be the same if $A=0$ instead of B.

Species III. Let $\frac{\mu}{h^2}>1$, and B negative, then by equation (2),

$$u=A\epsilon^{k\theta}-B\epsilon^{-k\theta}.$$

Putting $u=0$, we obtain as for Species I,

$$u=c\left\{\epsilon^{k(\theta-\alpha)}-\epsilon^{-k(\theta-\alpha)}\right\}.$$

Hence, when $\theta=\alpha$, $u=0$ or $r=\infty$. As θ increases r decreases, and when θ is infinite $r=0$. It is easily shewn that this curve has an asymptote parallel to OA, at a distance $\frac{1}{2ck}$.

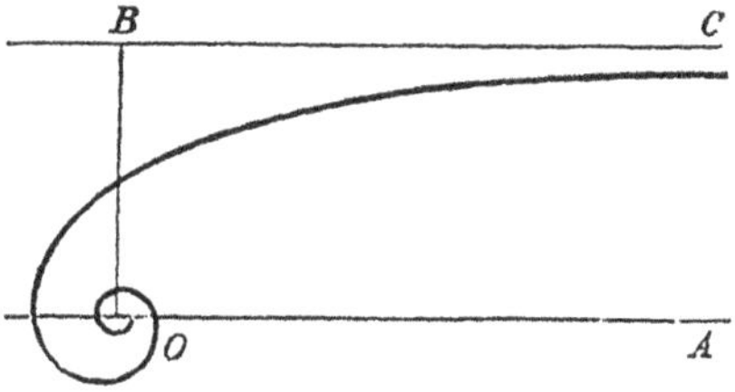

Species IV. Let $\frac{\mu}{h^2}=1$. Then equation (1) becomes

$$\frac{d^2u}{d\theta^2}=0,$$

the integral of which is

$$u=A(\theta-B),$$

the equation to the reciprocal spiral.

Species V. Let $\frac{\mu}{h^2}<1$, and let $1-\frac{\mu}{h^2}=k^2$, then by equation (1),

$$\frac{d^2u}{d\theta^2}+k^2u=0,$$

the integral of which is,

$$u = A \cos (k\theta + B);$$

$$\therefore \frac{du}{d\theta} = -Ak \sin (k\theta + B).$$

Let α be the value of θ corresponding to the apse,

$$\therefore k\alpha = -B;$$

$$\therefore u = A \cos k(\theta - \alpha),$$

which shews that $\frac{1}{A}$ is the apsidal distance. The asymptotes to this curve are easily found, and the curve is as represented in the fig.

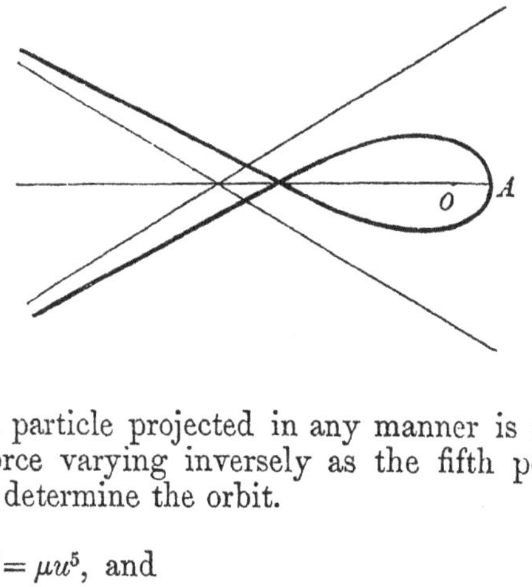

(16) A particle projected in any manner is acted on by a central force varying inversely as the fifth power of the distance, to determine the orbit.

Here, $P = \mu u^5$, and

$$\therefore \frac{d^2u}{d\theta^2} + u - \frac{\mu}{h^2} u^3 = 0;$$

$$\therefore h^2 \left\{ \left(\frac{du}{d\theta}\right)^2 + u^2 \right\} = v^2 = C + \frac{\mu u^4}{2}.$$

When $u = c$, $v = V$; $\therefore C = V^2 - \frac{\mu c^4}{2}$;

$$\therefore \frac{du^2}{d\theta^2} + u^2 = \frac{V^2}{h^2} - \frac{\mu c^4}{2h^2} + \frac{\mu u^4}{2h^2}.$$

But $h^2 = \dfrac{V^2 \sin^2\beta}{c^2}$;

$$\therefore \left(\frac{du}{d\theta}\right)^2 = \frac{\mu u^4 . c^2}{2V^2 \sin^2\beta} - u^2 + \frac{c^2}{\sin^2\beta} - \frac{\mu c^6}{2V^2 \sin^2\beta}.$$

This equation cannot be integrated in finite terms unless either the right hand member be a perfect square or $V^2 = \dfrac{\mu c^4}{2}$. The latter case is induced in the more general one of Example (8). In the former we must have

$$\frac{2\mu c^2}{V^2 \sin^2\beta}\left(\frac{c^2}{\sin^2\beta} - \frac{\mu c^6}{2V^2 \sin^2\beta}\right) = 1,$$

$$\text{or } 2\mu\left(V^2 - \frac{\mu c^4}{2}\right) = \frac{V^4}{c^4} \sin^4\beta,$$

from which β will be known.

Extracting now the square root, integrating and taking the initial line such that the constant $= 0$, we derive after the necessary reductions,

$$r = \pm \frac{\sqrt{(\mu)}}{VR \sin\beta} \frac{1 + \epsilon^{\theta\sqrt{2}}}{1 - \epsilon^{\theta\sqrt{2}}}.$$

Taking the upper sign we have a spiral having an interior and the lower an exterior asymptotic circle, the radius of this circle being in either case $\dfrac{\sqrt{(\mu)}}{VR \sin\beta}$.

(17) A particle is projected from an apse at the distance $\sqrt{(mh)}$, and is acted on by a central force $\dfrac{r}{m^2} + \dfrac{h^2}{r^3}$, h being

twice the area described in a unit of time; find the equation to the orbit and the time of describing a given angle.

Ans. $r^2 = \dfrac{mh}{1+\theta^2}$, $t = m \tan^{-1}\theta$.

(18) If a particle move about a center of force $\propto \dfrac{\mu}{r^2} + \dfrac{\nu}{r^3}$, shew that the equation to the orbit is generally of the form

$$r = \frac{a}{1 - e\cos(k\theta)}.$$

In the case when the projection takes place at an apse, the apsidal distance being $\dfrac{\mu}{h}$, and ν being $= h^2$, shew that the equation to the path is

$$r = \frac{2\mu h^2}{2h^3 + \mu^2\theta^2},$$

and that the time of describing an angle α is

$$\frac{1}{\alpha}\tan\theta\,(\theta + \tfrac{1}{2}\sin 2\theta) \text{ where } \tan\theta = \frac{\mu\alpha}{\sqrt{(2h^3)}}.$$

(19) If v be the velocity, and P the force at distance r in a central orbit, and if v', P', r' be similar quantities for the corresponding point of the locus of the foot of the perpendicular on the tangent, shew that

$$\frac{v^2}{Pr} + \frac{P'r'}{v'^2} = 2.$$

(20) A particle attached to one end of an elastic string moves on a smooth horizontal plane, the other end of the string being fixed to a point in the plane. If the path of the particle be a circle, shew that the periodic time $\propto \left(\dfrac{ra}{r-a}\right)^{\frac{1}{2}}$, a and r being the natural and stretched lengths of the string. If the orbit be nearly circular, find the angle between the apsides.

(21) A particle is projected in such a manner as to describe a reciprocal spiral whose equation is $\theta = \frac{a}{r}$; shew that the time of performing the nth revolution $= \frac{a^2}{2n(n-1)\pi\sqrt{(\mu)}}$.

(22) If P be a central force attracting a catenary, and p be the perpendicular on the tangent at any point from the center of force; then the force which would cause a particle to revolve in the curve formed by the catenary $\propto \frac{P}{p}$.

(23) Find the time in which a particle would move from the vertex to the end of the latus rectum of a parabola; and shew that if the velocity be there suddenly altered in the ratio $m : 1$ (m being < 1) the body will proceed to describe an ellipse, the excentricity of which $= (1 - 2m^2 + 2m^4)^{\frac{1}{2}}$.

(24) A spherical surface is described in space, having in its center a force varying inversely as the square of the distance; shew that if a particle be let fall from this surface and be projected in any direction at any moment of its descent with the velocity acquired, it will move in an ellipse, the major axis of which is equal to the radius of the sphere.

(25) If the Earth's orbit be taken an exact circle, and a comet be supposed to describe round the Sun a parabolic orbit in the same plane; shew that the comet cannot possibly continue within the Earth's orbit longer than the $\left(\frac{2}{3\pi}\right)^{th}$ part of a year.

(26) If a particle be projected about a center of force varying inversely as the square of the distance, with a velocity equal to n times the velocity in a circle at the same distance; the angle α between the axis major and this distance may be determined from the equation

$$\tan(\alpha - \beta) = (1 - n^2)\tan\beta.$$

(27) A particle describes a parabola about a center of force residing in a point in the circumference of a given ellipse the foci of which are in the circumference of the parabola; shew that the time of moving from one focus to the other is the same, at whatever point in the circumference of the ellipse the center of force is placed.

(28) A particle moves about a center of force, and its velocity at any point is inversely proportional to the distance from the center of force; shew that its path will be a logarithmic spiral.

(29) A particle is describing a curve about a center of force, and its velocity $\propto \frac{1}{r^n}$, find the law of force and the equation to the path.

$$\text{Ans.}\quad P \propto \frac{1}{r^{2n+1}},\quad \left(\frac{r}{a}\right)^{n-1} = \cos\{(n-1)\,\theta + \alpha\}.$$

(30) A particle is projected in any direction from one extremity of a straight line each particle of which attracts it by a force proportional to the distance, prove that the particle will pass through the other extremity.

(31) A particle projected in a given direction with a given velocity and attracted towards a given center of force has its velocity at every point to the velocity in a circle at the same distance as 1 to $\sqrt{(2)}$; find the orbit described, the position of the apse and the law of force.

$$\text{Ans.}\quad r = \sqrt{\left(\frac{\mu}{2h^2}\right)}\cos(\theta - \alpha),\quad P = \frac{\mu}{r^5}.$$

(32) A particle is projected from a given point with a given velocity and is acted on by a central force varying inversely as the square of the distance; shew that whatever be the direction of projection the center of the orbit described will lie on the surface of a certain sphere.

(33) Find the locus of the center of force that a cycloid may be described with uniform velocity, and find the law of force to the moving center.

(34) If a particle revolve in a circle radius r, about a center of force distant a from the center of the circle, shew that the time from distance r to the nearer apse is

$$\frac{2^{\frac{3}{2}} r^{\frac{1}{2}}}{\sqrt{(\phi)}\left(2-\frac{a^2}{r^2}\right)^{\frac{3}{2}}}\left\{\cos^{-1}\frac{a}{2r}-\frac{a}{2r}\sqrt{\left(1-\frac{a^2}{4r^2}\right)}\right\},$$

where ϕ is the initial force; and that the periodic time is

$$\frac{2\pi r^{\frac{3}{2}}}{(r-a)\sqrt{(\phi)}},$$

where ϕ is the force at the nearer apse.

(35) For any law of force shew how to find the orbit in which the velocity at each point is equal to that in a circle at the same distance. Shew that if $P \propto u^3$, the orbit is the logarithmic spiral.

(36) A particle describes an equilateral hyperbola about a center of force in the center, shew that an angle θ from the apsidal line is connected with the time t of its description by the formula

$$\sin 2\theta = \frac{\epsilon^{4\sqrt{(\mu)}.t}-1}{\epsilon^{4\sqrt{(\mu)}.t}+1}.$$

(37) If a number of particles, describing different circles in the same plane about a center of force $\propto D^{-3}$, start together from the same radius, find the curve in which they all lie when that which moves in the circle whose radius is a has completed a revolution.

(38) If the mth power of the periodic time be proportional to the nth power of the velocity in a circle, find the law of force in terms of the radius.

(39) If v be the velocity of a particle revolving in an ellipse about the center, v' its velocity when the direction of its

motion is at right angles to the former direction, the time of describing the intercepted arc $= \frac{1}{\sqrt{(\mu)}} \sin^{-1} \frac{vv'}{\mu ab}$.

(40) A particle revolves in a circle about a center of force in the center, the force $\propto \frac{1}{D^2}$; the absolute force is suddenly increased in the ratio of $m : 1$ when the particle is at any assigned point of its path, and when the particle arrives again at the same point the absolute force is again increased in the same ratio; shew that the path which the particle will describe is an ellipse whose excentricity $= \frac{m^2 - 1}{m^2}$.

(41) In a curve described by a particle under the action of a central force the angle between the radius vector and the tangent varies as the time. Find the curve and law of force.

(42) Shew that the apsidal angle is the same for different apsidal distances, only when the force is as some power of the distance.

(43) Given $P = \frac{2a^2h^2}{r^5} + \frac{h^2}{r^3}$, determine the path. Shew that in the particular case of the projection being made at distance a, and with velocity $= \frac{h}{a}\sqrt{(2)}$, the equation to the orbit is

$$r = a(1 + \theta).$$

(44) Force $= \frac{\mu}{r^4}$, and a particle is projected from an apse at distance a with velocity $= \sqrt{\left(\frac{2\mu}{3a^3}\right)}$, shew that the path is a cardioide, and that the periodic time is

$$\frac{3\pi}{4}\sqrt{\left(\frac{3a^5}{2\mu}\right)}.$$

(45) A particle is revolving in an ellipse about a center of force in the focus; supposing that every time the particle

arrives at the lower apse the absolute force is diminished in the ratio of 1 to $1-n$; find the excentricity of the elliptic orbit after p revolutions, the original excentricity being e.

Ans. $\dfrac{1+e}{(1-n)^p}-1.$

(46) A particle describes a circular orbit about a center of force situated in the center of the circle; prove that the form of the orbit will be stable or unstable according as the value of $\dfrac{d\log P}{d\log u}$, for $u=a$, is less or not less than 3, P being the central force, u the reciprocal of the radius vector, and $\dfrac{1}{a}$ the radius of the circle.

(47) If the equation for determining the apsidal distances in a central orbit contain the factor $(u-a)^p$, shew that $u=a$ cannot correspond to an apse unless p be of one of the forms $4m+2$ or $\dfrac{4m+2}{2n+1}$. If the factor $u-a$ occur twice, then a will be a root of the equation

$$\phi(u)-h^2u^3=0,$$

where $\phi(u)$ is the central force.

CHAPTER VI.

ELLIPTIC MOTION.

108. IN this chapter we propose to deduce from the results of the last some of the properties of Elliptic and Parabolic Orbits described about a center of force in the focus. This is a problem of great interest, as it has been proved by actual observation that the orbits of planets and comets are in general (neglecting the small effects of disturbing forces) ellipses either very slightly excentric, or so much so as to be scarcely distinguishable from parabolas. There are, it is true, some comets whose orbits are moderately excentric ellipses, and some whose orbits are hyperbolas; but, as the problem in this case becomes very complicated, and the approximate methods which we will here employ are inapplicable to their motions, it has been considered advisable to omit the consideration of such cases.

109. In order to understand what follows it will be necessary to premise a few definitions.

Suppose APA' to be an elliptic orbit described about a center of force in the focus S. Also suppose P the position of the particle at any time t. Draw PM perpendicular to the major axis ACA', and produce it to cut the auxiliary circle in the point Q. Let C be the center of the curves. Join CQ.

When the moving particle is at A, the nearest point of the orbit to S, it is said to be in perihelion.

The angle ASP, or the excess of the particle's longitude over that of the perihelion, is called the true anomaly. Let us denote it by θ.

The angle ACQ is called the excentric anomaly, and is generally denoted by u. And if $\frac{2\pi}{n}$ is the time of a complete revolution, nt is the circular measure of an imaginary angle called the mean anomaly; it would evidently be the true anomaly if the particle's angular velocity about S were uniform.

110. It is easy from known properties of the ellipse to deduce relations between the mean and excentric, and also between the true and excentric, anomalies; and this we proceed to do.

PROB. *To find the relation between the mean and excentric anomalies.*

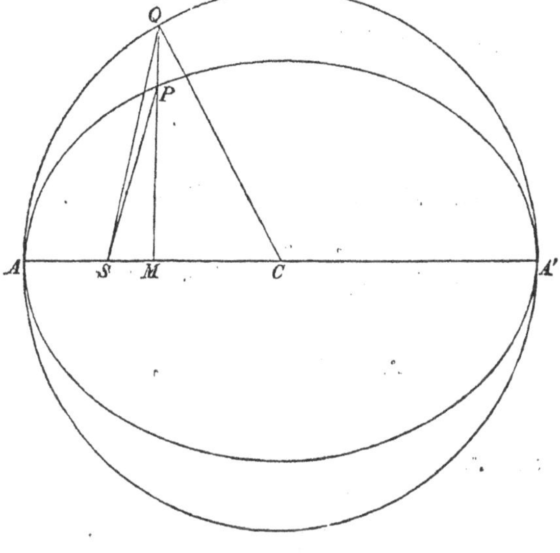

In the figure QCA is the excentric anomaly, and the mean anomaly is evidently to 2π as the area PSA is to the whole area of the elliptic orbit (Arts. 103, 109), or as area QSA to that of the auxiliary circle.

Now area QSA = area QCA − area QCS

$$= \tfrac{1}{2} a^2 u - \tfrac{1}{2} a \,.\, ae \,.\, \sin u$$

(a being the semi-major axis of the orbit and e its excentricity)

$$=\frac{a^2}{2}(u-e\sin u).$$

$$\text{Hence }\frac{nt}{2\pi}=\frac{\frac{a^2}{2}(u-e\sin u)}{\pi a^2};$$

$$\text{or } nt=u-e\sin u.$$

111. PROB. *To find the relation between the true and excentric anomalies.*

We have (by Conics)

$$SP=\frac{a(1-e^2)}{1+e\cos\theta}.$$

$$\text{But } SP=a-eCM=a(1-e\cos u).$$

$$\text{Hence }\frac{1-e^2}{1+e\cos\theta}=1-e\cos u;$$

$$\therefore \cos\theta=\frac{\cos u-e}{1-e\cos u},$$

$$\text{and } \therefore \tan\frac{\theta}{2}=\sqrt{\left(\frac{1-\cos\theta}{1+\cos\theta}\right)}$$

$$=\sqrt{\left(\frac{1-e\cos u-\cos u+e}{1-e\cos u+\cos u-e}\right)}$$

$$=\sqrt{\left\{\frac{(1+e)(1-\cos u)}{(1-e)(1+\cos u)}\right\}}$$

$$=\sqrt{\left(\frac{1+e}{1-e}\right)}.\tan\frac{u}{2}.$$

These two equations which we have just found are sufficient for the solution of our problem, but they are sometimes derived in the following manner.

112. The direct problem in elliptic motion is

To find the time of motion of a planet or comet through any portion of its elliptic orbit.

The equation to the orbit gives us

$$r = \frac{a(1-e^2)}{1+e\cos\theta}.$$

And from the description of equal areas in equal times, we have,

$$\frac{d\theta}{dt} = \frac{h}{r^2} = \frac{\sqrt{\{\mu a(1-e^2)\}}}{r^2}.$$

From the latter we have

$$\frac{dt}{d\theta} = \frac{a^{\frac{3}{2}}(1-e^2)^{\frac{3}{2}}}{\sqrt{(\mu)}} \frac{1}{(1+e\cos\theta)^2} = \frac{(1-e^2)^{\frac{3}{2}}}{n} \frac{1}{(1+e\cos\theta)^2},$$

if $\frac{2\pi}{n}$ is the period of revolution.

Hence if t_1 be the time of describing an arc measured by θ_1 from perihelion,

$$nt_1 = (1-e^2)^{\frac{3}{2}} \int_0^{\theta_1} \frac{d\theta}{(1+e\cos\theta)^2}, \text{ or } \left(\text{remarking that}\right.$$

$$\left.\cos\theta = \cos^2\frac{\theta}{2} - \sin^2\frac{\theta}{2}, \text{ and } 1 = \cos^2\frac{\theta}{2} + \sin^2\frac{\theta}{2}\right)$$

$$= (1-e^2)^{\frac{3}{2}} \int_0^{\theta_1} \frac{d\theta}{\left\{(1+e)\cos^2\frac{\theta}{2} + (1-e)\sin^2\frac{\theta}{2}\right\}^2}$$

$$= 2(1-e^2)^{\frac{3}{2}} \int_0^{\theta_1} \frac{\sec^2\frac{\theta}{2} \frac{d\left(\tan\frac{\theta}{2}\right)}{d\theta} . d\theta}{\left\{(1+e) + (1-e)\tan^2\frac{\theta}{2}\right\}^2}.$$

To simplify this, let us put

$$\tan\frac{\theta}{2} = \sqrt{\left(\frac{1+e}{1-e}\right)} \cdot \tan\frac{\phi}{2} \quad \ldots\ldots\ldots\ldots\ldots (1),$$

(an assumption which will evidently conduct us to a formula already proved, as it is clear that ϕ will represent the excentric anomaly).

We have

$$nt_1 = 2(1-e^2)^{\frac{3}{2}} \int_0^{\phi_1} \frac{\left(1 + \frac{1+e}{1-e}\tan^2\frac{\phi}{2}\right)}{(1+e)^2 \sec^4\frac{\phi}{2}} \frac{d}{d\phi}\left\{\sqrt{\left(\frac{1+e}{1-e}\right)}\tan\frac{\phi}{2}\right\} d\phi$$

$$= \int_0^{\phi_1} \left\{(1-e)\cos^2\frac{\phi}{2} + (1+e)\sin^2\frac{\phi}{2}\right\} d\phi$$

$$= \int_0^{\phi_1} (1 - e\cos\phi)\, d\phi$$

$$= \phi_1 - e\sin\phi_1 \quad \ldots\ldots\ldots\ldots\ldots\ldots\ldots\ldots (2).$$

When θ_1 is given we must calculate ϕ_1 by means of (1), and thence t_1 by means of (2).

Since (1) and (2) give us the time of passing through an arc from perihelion subtending any angle θ_1 at the focus, it is evident that we have now the means of finding the time of describing any given portion of the orbit, and have thus the complete solution of the direct problem.

113. The inverse problem, which is far the most important, is to find the values of θ and r as functions of t, so that the direction in which a planet is, and its distance, may be determined for any given time. This generally goes by the name of Kepler's Problem.

114. Before entering on the systematic development of u, r and θ in terms of t from our equations, it may be useful to remark that if e be so small that higher terms than its

square may be neglected, we may easily obtain developments correct to the first three terms.

Thus $u = nt + e \sin u$

$$= nt + e \sin (nt + e \sin nt) \text{ nearly,}$$

$$= nt + e \sin nt + \frac{e^2}{2} \sin 2nt.$$

Also $\frac{r}{a} = 1 - e \cos u$

$$= 1 - e \cos (nt + e \sin u)$$

$$= 1 - e \cos nt + \frac{e^2}{2} (1 - \cos 2nt).$$

$$\text{And } r^2 . \frac{d\theta}{dt} = \sqrt{\{\mu a (1 - e^2)\}},$$

which may be written

$$\frac{a^2 (1 - e^2)^2}{(1 + e \cos \theta)^2} \cdot \frac{d\theta}{dt} = na^2 (1 - e^2)^{\frac{1}{2}},$$

$$\text{or } (1 - e^2)^{\frac{3}{2}} (1 + e \cos \theta)^{-2} \frac{d\theta}{dt} = n.$$

Keeping powers of e lower than the third

$$(1 - 2 e \cos \theta + \frac{3}{2} e^2 \cos 2\theta) \frac{d\theta}{dt} = n,$$

$$\text{or } nt = \theta - 2e \sin \theta + \frac{3}{4} e^2 \sin 2\theta;$$

whence $\theta = nt + 2e \sin \theta - \frac{3}{4} e^2 \sin 2\theta$

$$= nt + 2e \sin (nt + 2e \sin nt) - \frac{3}{4} e^2 \sin 2nt$$

$$= nt + 2e \sin nt + 4e^2 \cos nt \sin nt - \frac{3}{4} e^2 \sin 2nt$$

$$= nt + 2e \sin nt + \frac{5}{4} e^2 \sin 2nt.$$

115. PROB. *To find r and θ as functions of t from the equations*

$$r = a(1 - e \cos u) \ldots\ldots\ldots\ldots (1);$$

$$\tan \frac{\theta}{2} = \sqrt{\left(\frac{1+e}{1-e}\right)} \cdot \tan \frac{u}{2} \ldots\ldots\ldots\ldots (2);$$

$$nt = u - e \sin u \ldots\ldots\ldots\ldots (3).$$

These equations evidently give r, θ, and t directly for any assigned value of u, but this is of little value in practice. The method of solution which we proceed to give is that of Lagrange, and the general principle of it is this—

We can develop θ from equation (2) in a series ascending by powers of a small function of e, the coefficients of these powers being u and the sines of multiples of u. Now by Lagrange's theorem we may from equation (3) express u, $1 - e \cos u$, $\sin u$, $\sin 2u$, &c. in series ascending by powers of e, and whose coefficients are sines or cosines of multiples of nt. Hence by substituting these values in equation (1) and in the development of (2), we have r and θ expressed in series whose terms rapidly decrease, and whose coefficients are sines or cosines of multiples of nt. And this is the complete solution of the problem.

116. *To express the true anomaly as a function of the excentric.*

Substituting in (2) the exponential expressions for the tangents, we have

$$\frac{\epsilon^{\frac{\theta}{2}\sqrt{(-1)}} - \epsilon^{-\frac{\theta}{2}\sqrt{(-1)}}}{\epsilon^{\frac{\theta}{2}\sqrt{(-1)}} + \epsilon^{-\frac{\theta}{2}\sqrt{(-1)}}} = \sqrt{\left(\frac{1+e}{1-e}\right)} \cdot \frac{\epsilon^{\frac{u}{2}\sqrt{(-1)}} - \epsilon^{-\frac{u}{2}\sqrt{(-1)}}}{\epsilon^{\frac{u}{2}\sqrt{(-1)}} + \epsilon^{-\frac{u}{2}\sqrt{(-1)}}},$$

whence

$$\epsilon^{\theta\sqrt{(-1)}} = \frac{\epsilon^{u\sqrt{(-1)}}\{\sqrt{(1+e)}+\sqrt{(1-e)}\}+\{\sqrt{(1-e)}-\sqrt{(1+e)}\}}{\epsilon^{u\sqrt{(-1)}}\{\sqrt{(1-e)}-\sqrt{(1+e)}\}+\{\sqrt{(1-e)}+\sqrt{(1+e)}\}};$$

or, putting $\lambda = \dfrac{\sqrt{(1+e)}-\sqrt{(1-e)}}{\sqrt{(1+e)}+\sqrt{(1-e)}} = \dfrac{e}{1+\sqrt{(1-e^2)}}$,

$$\epsilon^{\theta\sqrt{(-1)}} = \epsilon^{u\sqrt{(-1)}} \cdot \frac{1-\lambda\epsilon^{-u\sqrt{(-1)}}}{1-\lambda\epsilon^{u\sqrt{(-1)}}}.$$

Taking the logarithm of each side and dividing by $\sqrt{(-1)}$,

$$\theta = u + \frac{\lambda}{\sqrt{(-1)}}\{\epsilon^{u\sqrt{(-1)}} - \epsilon^{-u\sqrt{(-1)}}\} + \frac{\lambda^2}{2\sqrt{(-1)}}\{\epsilon^{2u\sqrt{(-1)}} - \epsilon^{-2u\sqrt{(-1)}}\} + \ldots$$

$$= u + 2\left(\lambda \sin u + \frac{\lambda^2}{2}\sin 2u + \frac{\lambda^3}{3}\sin 3u + \&c.\right) \ldots\ldots\ldots\ldots (4).$$

117. *To develop u in terms of t.*

If we have

$$y = z + x\phi(y) \ldots\ldots\ldots\ldots\ldots\ldots\ldots\ldots (5),$$

we know (TODHUNTER, *Diff. Calc.* Art. 138) that

$$f(y) = f(z) + x\phi(z)f'(z) + \frac{x^2}{\lfloor 2}\frac{d}{dz}\{\overline{\phi(z)}|^2 f'(z)\}$$

$$+ \frac{x^3}{\lfloor 3}\left(\frac{d}{dz}\right)^2\{\overline{\phi(z)}|^3 f'(z)\} + \&c. \ldots\ldots\ldots (6).$$

Now equation (3) may be put in the form

$$u = nt + e\sin u,$$

which is identical in form with (5)

$$\{\text{if } y = u,\ z = nt,\ x = e, \text{ and } \phi(y) = \sin y\}.$$

Also as it is the development of u that we require,

$$f(u) = u, \text{ and } f'(u) = 1. \text{ Hence, by (6),}$$

$$y = z + x \sin z + \frac{x^2}{1.2} \cdot \frac{d}{dz} (\sin^2 z) + \frac{x^3}{1.2.3} \left(\frac{d}{dz}\right)^2 (\sin^3 z) + \&c.;$$

and substituting for the powers of $\sin z$ their corresponding expressions in sines and cosines of multiple angles;

$$y = z + x \sin z + \frac{x^2}{1.2} \frac{d}{dz} \left(\frac{1 - \cos 2z}{2}\right) + \frac{x^3}{1.2.3} \left(\frac{d}{dz}\right)^2 \left(\frac{3 \sin z - \sin 3z}{4}\right)$$

$$+ \frac{x^4}{1.2.3.4} \left(\frac{d}{dz}\right)^3 \left(\frac{3 - 4 \cos 2z + \cos 4z}{8}\right) + \&c.$$

$$= z + x \sin z + \frac{x^2}{2} \sin 2z + \frac{x^3}{8} (3 \sin 3z - \sin z) + \ldots\ldots$$

or, substituting for x, y, z their values as above,

$$u = nt + e \sin nt + \frac{e^2}{2} \sin 2nt + \frac{e^3}{8} (3 \sin 3nt - \sin nt)$$

$$+ \frac{e^4}{6} (2 \sin 4nt - \sin 2nt) + \&c. \ldots\ldots (7).$$

To develop $\sin u$, we recur to equation (3), which gives after the elimination of u by means of (7),

$$\sin u = \sin nt + \frac{e}{2} \sin 2nt + \frac{e^2}{8} (3 \sin 3nt - \sin nt) + \&c. \ldots (8).$$

By the application of Lagrange's theorem to equation (3), it is easy to deduce the following expressions:

$$\sin 2u = \sin 2nt + e (\sin 3nt - \sin nt) + e^2 (\sin 4nt - \sin 2nt)$$

$$+ \frac{e^3}{24} (4 \sin nt - 27 \sin 3nt + 25 \sin 5nt) + \&c.$$

$$\sin 3u = \sin 3nt + \frac{3e}{2}(\sin 4nt - \sin 2nt)$$

$$+\frac{e^2}{8}(15 \sin 5nt - 18 \sin 3nt + 3 \sin nt) + \&c.$$

$$\&c. = \&c.$$

Substituting these values in (4) we obtain the value of θ, containing however the quantity λ. If we take as its approximate value $\frac{e}{2}+\frac{e^3}{8}$, and make the requisite substitutions, we obtain

$$\theta = nt + (2e - \frac{1}{4}e^3)\sin nt + \frac{5}{4}e^2 \sin 2nt + \frac{13}{12}e^3 \sin 3nt + \ldots\ldots$$

which is correct as far as e^3.

118. In proceeding farther with the development, it becomes necessary to expand λ and its powers in series ascending by powers of e. This is readily done as follows.

We have

$$\lambda = \frac{e}{1+\sqrt{(1-e^2)}} = \frac{e}{E} \text{ suppose.}$$

Hence, $E = 2 - \frac{e^2}{E}$ from which by Lagrange's Theorem,

$$E^{-p} = \frac{1}{2^p} + \frac{p}{2^{p+2}}e^2 + \frac{p.(p+3)}{2.2^{p+4}}e^4 + \&c.;$$

and thus the value of $\lambda^p = e^p E^{-p}$ is known.

The correct value of θ to the fifth power of e is thus found to be

$$nt + 2e \sin nt + \frac{5e^2}{4} \sin 2nt + \frac{e^3}{2^2 . 3} (13 \sin 3nt - 3 \sin nt)$$

$$+ \frac{e^4}{2^5 . 3} (103 \sin 4\, nt - 44 \sin 2nt)$$

$$+ \frac{e^5}{2^6 . 3 . 5} (1097 \sin 5nt - 645 \sin 3nt + 50 \sin nt).$$

119. *To develop r in terms of t.*

From (1) it is evident that all we have to do is by Lagrange's Theorem to develop $1 - e \cos u$ as a function of t from $nt = u - e \sin u$.

To develop $(1 - e \cos u)$ *in terms of t.*

Here $f(y) = 1 - e \cos y$,

$$f'(y) = e \sin y;$$

and the form of ϕ is the same as before;

$$\therefore\ 1 - e \cos y = (1 - e \cos z) + x \sin z\, (e \sin z)$$

$$+ \frac{x^2}{1 . 2} \frac{d}{dz} (\sin^2 z . e \sin z) + \ldots\ldots\ldots$$

Hence as before, substituting for the powers of sines the equivalent expressions in sines and cosines of multiple arcs, differentiating, and substituting u for y, nt for z, and e for x, we have

$$1 - e \cos u = \frac{r}{a} = 1 - e \cos nt + \frac{e^2}{2} (1 - \cos 2nt)$$

$$+ \frac{e^3}{8} (3 \cos nt - 3 \cos 3nt)$$

$$+ \frac{e^4}{3} (\cos 2nt - \cos 4nt) + \&c.$$

which gives the radius vector in terms of the time.

120. In the case of parabolic motion the above methods are not applicable; but a much simpler one is.

PROP. *To find the time of describing any arc of a parabola from the vertex; the center of force being in the focus.*

The equation to the curve is

$$r = d\sec^2\frac{\theta}{2}, \text{ where } d \text{ is the perihelion distance.}$$

And the dynamical condition gives

$$r^2\frac{d\theta}{dt} = h = \sqrt{(2\mu d)};$$

$$t_1 = \frac{d^{\frac{3}{2}}}{\sqrt{(2\mu)}}\int_0^{\theta_1}\sec^4\frac{\theta}{2}\,d\theta$$

$$= \frac{\sqrt{(2)}\,.\,d^{\frac{3}{2}}}{\sqrt{(\mu)}}\int_0^{\theta_1}\left(1+\tan^2\frac{\theta}{2}\right).\frac{d\left(\tan\frac{\theta}{2}\right)}{d\theta}.\,d\theta$$

$$= \frac{\sqrt{(2)}\,.\,d^{\frac{3}{2}}}{\sqrt{(\mu)}}\left(\tan\frac{\theta_1}{2}+\frac{1}{3}.\tan^3\frac{\theta_1}{2}\right);$$

$$\text{or } nt_1 = \left(\tan\frac{\theta_1}{2}+\frac{1}{3}\tan^3\frac{\theta_1}{2}\right), \text{ if } \sqrt{\left(\frac{\mu}{2d^3}\right)} = n;$$

which is the expression required. From this it is evidently easy to calculate the time through any arc of the orbit.

The inverse problem of parabolic motion would require the solution of the cubic equation just found for $\tan\frac{\theta}{2}$ in terms of the time. This however is easily obviated by the formation of a table in which corresponding values of t and $\frac{\theta}{2}$ are calculated on the supposition that $n=1$. If we wish then to solve the inverse problem all we have to do is to find the value of θ corresponding to the number nt. This will be the true anomaly required, and the same table will of course apply to any parabolic orbit.

121. If the orbit be not parabolic, but elliptic and of very great excentricity, the following direct problem is sometimes of use.

PROB. *To find the place of a comet at a given time in a very excentric elliptic orbit.*

We have $\dfrac{dt}{d\theta} = \dfrac{a^{\frac{3}{2}}(1-e^2)^{\frac{3}{2}}}{\sqrt{(\mu)}} \cdot \dfrac{1}{(1+e\cos\theta)^2}$.

Let D be the perihelion distance, $D = a(1-e)$;

$$\frac{dt}{d\theta} = \frac{D^{\frac{3}{2}}(1+e)^{\frac{3}{2}}}{\sqrt{(\mu)}} \cdot \frac{\sec^4\dfrac{\theta}{2}}{\left\{(1+e)+(1-e)\tan^2\dfrac{\theta}{2}\right\}^2}$$

$$= \frac{D^{\frac{3}{2}}}{\sqrt{\{\mu(1+e)\}}} \sec^4\frac{\theta}{2}\left\{1+\frac{1-e}{1+e}\cdot\tan^2\frac{\theta}{2}\right\}^{-2}.$$

Expanding in powers of $(1-e)$, and neglecting higher powers of $(1-e)$ than the first; since $e=1$ nearly,

$$nt_1 = \frac{1}{2}\left(1-\frac{1-e}{2}\right)^{-\frac{1}{2}} \int_0^{\theta_1} \sec^4\frac{\theta}{2}\left\{1-(1-e)\tan^2\frac{\theta}{2}\right\} d\theta$$

$$= \int_0^{\theta_1} \frac{d\tan\dfrac{\theta}{2}}{d\theta}\left\{1+\tan^2\frac{\theta}{2}+(1-e)\left(\frac{1}{4}-\frac{3}{4}\tan^2\frac{\theta}{2}-\tan^4\frac{\theta}{2}\right)\right\} d\theta;$$

$$(1) \; \therefore nt_1 = \tan\frac{\theta_1}{2}+\frac{1}{3}\tan^3\frac{\theta_1}{2}+(1-e)\left(\frac{1}{4}\tan\frac{\theta_1}{2}-\frac{1}{4}\tan^3\frac{\theta_1}{2}-\frac{1}{5}\tan^5\frac{\theta_1}{2}\right).$$

The following is a convenient method of calculating the value of θ_1 for a given value of t_1.

Suppose $\bar{\theta}$ is the true anomaly at time t_1, of a comet moving in a parabolic orbit of which D is the perihelion distance; then

$$nt_1 = \tan\frac{\bar{\theta}}{2} + \frac{1}{3}\tan^3\frac{\bar{\theta}}{2} \ldots\ldots\ldots\ldots (2).$$

Let $\theta_1 = \bar{\theta} + x$, substitute in equation (1) and since x is very small, expand in powers of x by Taylor's Theorem, we have approximately

$$nt_1 = \tan\frac{\bar{\theta}}{2} + \frac{1}{3}\tan^3\frac{\bar{\theta}}{2} + \frac{1}{2}x\sec^4\frac{\bar{\theta}}{2} + \ldots\ldots$$

$$+\frac{1}{4}(1-e)\tan\frac{\bar{\theta}}{2}\left(1-\tan^2\frac{\bar{\theta}}{2} - \frac{4}{5}\tan^4\frac{\bar{\theta}}{2}\right).$$

From this by means of (2), we obtain

$$x = \frac{1}{10}(1-e)\tan\frac{\bar{\theta}}{2}\left(4 - 3\cos^2\frac{\bar{\theta}}{2} - 6\cos^4\frac{\bar{\theta}}{2}\right).$$

To make use of this formula, there must be added to the table before mentioned, a column giving the values of $\frac{x}{1-e}$ corresponding to those of $\bar{\theta}$. Taking then any value of t, we seek in the second column the value of $\bar{\theta}$ for the number nt, then the value of $\frac{x}{1-e}$ for the value of $\bar{\theta}$ so found.

And as the orbit is known, $1-e$ is known, hence x and $\bar{\theta}$ are known in terms of t, and the true anomaly or

$$\bar{\theta} + x \text{ is known.}$$

122. Remark. In all that precedes we have supposed for simplicity that the angle θ which determines the position of the particle is measured from perihelion, and that $\theta = 0$, $t = 0$ together. This is not usually the case, but let θ' be the *longitude* of the particle at time t, π that of the perihelion, ϵ the longitude of the particle at time $t = 0$, or the *epoch* as it is generally called; then at time t the mean longitude is

evidently $nt+\epsilon$, and the mean anomaly $nt+\epsilon-\pi$. Hence by our previous results

$$\theta'-\pi = nt+\epsilon-\pi+2e\sin(nt+\epsilon-\pi)$$
$$+\frac{5}{4}e^2\sin 2(nt+\epsilon-\pi)+\&c.$$

$$r = a\{1-e\cos(nt+\epsilon-\pi)+\ldots\},$$

which are the formulæ in general use, θ' being, as before observed, the true longitude at time t.

123. Prop. *The time through any arc of a parabolic orbit described about the focus, may be expressed in terms of the chord and extreme radii vectores of the arc.*

Let $r_1, \theta_1, r_2, \theta_2$ be the co-ordinates of the extreme points, c the chord, of the arc. Then T being the required time, we have (Art. 120).

$$nT = \tan\frac{\theta_2}{2}-\tan\frac{\theta_1}{2}+\frac{1}{3}\left(\tan^3\frac{\theta_2}{2}-\tan^3\frac{\theta_1}{2}\right),$$

or as we may write for simplicity

$$= t_2-t_1+\frac{1}{3}(t_2^3-t_1^3)$$

$$= \frac{1}{3}(t_2-t_1)(3+t_1^2+t_2^2+t_1t_2)$$

$$= \frac{1}{3}(t_2-t_1)\{(1+t_1^2)+(1+t_2^2)+(1+t_1t_2)\}.$$

$$\text{Now } 1+t_1t_2 = \frac{\cos\frac{\theta_2-\theta_1}{2}}{\cos\frac{\theta_1}{2}\cos\frac{\theta_2}{2}},$$

and in the triangle whose base is c, sides $r_1 r_2$, and vertical angle $\theta_2-\theta_1$, we have by Trigonometry

$$\cos\frac{\theta_2-\theta_1}{2}=\sqrt{\left\{\frac{s.(s-c)}{r_1r_2}\right\}}\quad\left(\text{where } s=\frac{r_1+r_2+c}{2}\right).$$

$$\text{Also, } r_1=a\sec^2\frac{\theta_1}{2},\quad r_2=a\sec^2\frac{\theta_2}{2}.$$

$$\text{Hence, } 1+t_1t_2=\frac{1}{a}\sqrt{\{s.(s-c)\}}.$$

$$\text{And } t_2-t_1=\sqrt{\{1+t_2^2+1+t_1^2-2(1+t_1t_2)\}}$$

$$=\sqrt{\frac{1}{a}[r_2+r_1-2\sqrt{\{s.(s-c)\}}]}$$

$$=\sqrt{\frac{1}{a}[2s-c-2\sqrt{\{s.(s-c)\}}]}$$

$$=\frac{1}{\sqrt{(a)}}\{\sqrt{(s)}-\sqrt{(s-c)}\}.$$

$$\text{Also } 1+t_1^2+1+t_2^2+1+t_1t_2=\frac{1}{a}[r_1+r_2+\sqrt{\{s.(s-c)\}}]$$

$$=\frac{1}{a}[2s-c+\sqrt{\{s.(s-c)\}}].$$

$$\text{Hence } nT=\frac{1}{3a^{\frac{3}{2}}}\{\sqrt{(s)}-\sqrt{(s-c)}\}\{s+\sqrt{(s)}.\sqrt{(s-c)}+(s-c)\}$$

$$=\frac{1}{3a^{\frac{3}{2}}}\{s^{\frac{3}{2}}-(s-c)^{\frac{3}{2}}\}.$$

$$\text{But } n=\sqrt{\left(\frac{\mu}{2a^3}\right)}.\quad\text{Art. (120).}$$

Substituting this and the values of s, $(s-c)$, we have

$$T=\frac{1}{6\sqrt{(\mu)}}\{(r_1+r_2+c)^{\frac{3}{2}}-(r_1+r_2-c)^{\frac{3}{2}}\}.$$

In this investigation we have supposed the arc not to include the perihelion; if it should do so we must take the *sum* of the radicals as the value of T.

124. It may be shewn in a similar manner that the time of describing about the focus an arc of an ellipse or hyperbola whose chord and extreme radii vectores are given, may be expressed in terms of these quantities and the axis major alone. For the proof we must refer to the "Mécanique Céleste," or to Pontécoulant's "Système du Monde."

It may also be shewn in much the same manner that the ratio of the area described in a given time, to that of the triangle formed by the chord and extreme radii vectores, may be expressed independently of the parameter of the path.

EXAMPLES.

(1) If the perihelion distance of a comet's orbit be $\frac{1}{3}$ of the radius of the Earth's orbit supposed circular, find the number of days the comet will remain within the Earth's orbit.

(2) If a comet describe 90° from perihelion in 100 days, compare its perihelion distance with the distance of a planet which describes its circular orbit in 942 days.

(3) Shew how to divide a planet's elliptic orbit by a diameter, so that the times of describing the two parts are as $n : 1$, and find in what cases only one such line can be drawn.

(4) In the case of planets and comets prove the following formulæ, the letters being the same as in the text,

$$r\frac{d\theta}{du} = a\sqrt{(1-e^2)};$$

$$\frac{r}{a}\sin\theta = \frac{\sqrt{(1-e^2)}}{e}(u-nt);$$

$$\log\frac{r}{a} = -\log(1+\lambda^2)$$
$$-2\left(\lambda\cos u + \tfrac{1}{2}\lambda^2\cos 2u + \tfrac{1}{3}\lambda^3\cos 3u + \&c.\right)$$

(5) If Pp, Qq be chords parallel to the axis major of an elliptic orbit, shew that the difference of the times through the arcs PQ, pq varies as the distance between the chords.

(6) If a comet whose orbit is inclined to the plane of the ecliptic were observed to pass over the Sun's disc and three months after to strike the planet Mars, determine its distance from the Earth at the first observation, the Earth and Mars describing about the Sun circles whose radii are as 2 : 3.

(7) Shew that the arithmetic mean of the distances of a planet from the Sun, at equal indefinitely small intervals of time, is

$$a\left(1+\frac{e^2}{2}\right).$$

(8) The time through an arc of a parabolic orbit bounded by a focal chord $\propto$ (chord)$^{\frac{3}{2}}$.

(9) If a circle be described passing through the focus and vertex of a parabolic orbit, and also through the position of the moving particle at each instant, shew that its center describes with uniform velocity a straight line bisecting at right angles the perihelion distance.

(10) Shew that the velocity of a comet perpendicular to the major axis varies inversely as its radius vector.

(11) D_1, D_2 being two distances of a comet, on opposite sides of perihelion, including a known angle, shew that the position of perihelion may be found from the equation

$$\frac{\sqrt{D_1}-\sqrt{D_2}}{\sqrt{D_1}+\sqrt{D_2}} = \tan\tfrac{1}{4}\,(\text{sum of true anomalies})\,.\,\tan\tfrac{1}{4}\,(\text{difference}).$$

(12) In what point of all conic sections is the paracentric velocity a maximum? Shew that in such a case the velocity is to that in a circle at the same distance as the distance is to the perpendicular on the tangent.

(13) In an elliptic orbit find the relation between the mean angular velocity about the center of force and the angular velocity about the other focus, and thence shew that when e is small the latter is nearly uniform.

(14) If α, β be the greatest and least angular velocities in an ellipse about the focus, the *mean* angular velocity is

$$\frac{2\sqrt[4]{(\alpha^3\beta^3)}}{\sqrt{(\alpha)}+\sqrt{(\beta)}}.$$

(15) Find the maximum value of $\theta - nt$ in an elliptic orbit, and develop it in powers of e, shewing that it cannot contain even powers.

If E' be this quantity

$$E' = 2e + \frac{11e^3}{3 \cdot 2^4} + \frac{599e^5}{5 \cdot 2^{10}} + \&c.$$

CHAPTER VII.

CONSTRAINED MOTION.

125. We come now to the case of the motion of a particle subject to the action not only of given forces, but of undetermined pressures or tensions. Such cases occur when a particle is attached to a fixed point by means of a rod or string, or when it is forced to move on a curve or surface.

In applying to a problem of this kind the general equations of motion of a free particle, we must assume directions and intensities for the unknown forces, treating them then as known, and it will always be found that the geometrical circumstances of the motion will furnish the requisite number of additional equations for the determination of all the unknown quantities in terms of the time.

One case of this kind has been already treated of (Art. 38), namely, that of a particle moving on an inclined plane under the action of gravity. There the undetermined force is the pressure on the plane, which however is evidently constant, and equal to the resolved part of the particle's weight perpendicular to the plane.

Little is known about the laws of dynamical friction, and therefore the few investigations which will be given of motion on a rough curve or surface are scarce of any importance.

126. The simplest case is,

Prop. *A particle is constrained to move on a given smooth plane curve under the action of given forces in the plane of the curve, to determine the motion.*

Taking rectangular axes in this plane, the forces may be resolved into two, X and Y respectively parallel to the axes of x and y. In addition there will be the force R, the mutual pressure between the curve and particle, which evidently acts in the normal to the curve since the latter is smooth.

Let P be the position of the particle at the time t; and let

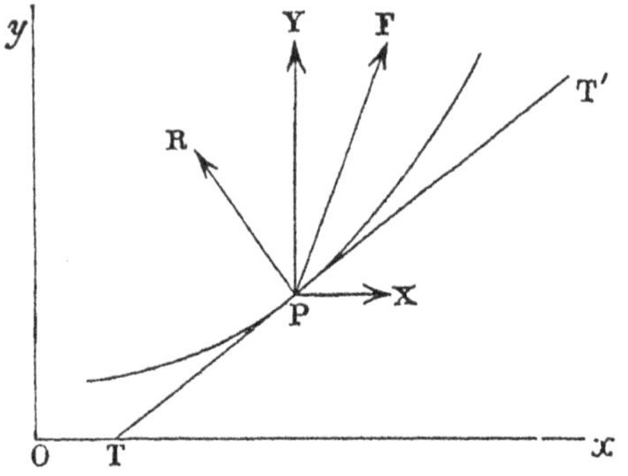

the forces X, Y, R, act as in the figure. Draw PT a tangent at P. Then if $PTx = \theta$, we have

$$\sin\theta = \frac{dy}{ds}, \quad \cos\theta = \frac{dx}{ds}.$$

Now if m be the mass of the particle, the equations of motion are

$$m\frac{d^2x}{dt^2} = mX - R\sin\theta = mX - R\frac{dy}{ds} \quad \text{.........} \quad (1),$$

$$m\frac{d^2y}{dt^2} = mY + R\cos\theta = mY + R\frac{dx}{ds} \quad \text{.........} \quad (2).$$

These two equations together with the given equation to the curve, are sufficient to determine the motion completely.

To eliminate R, multiply (1) by $\frac{dx}{dt}$ and (2) by $\frac{dy}{dt}$, and add. We thus obtain,

$$\left(\text{since } \frac{dy}{ds}\frac{dx}{dt} = \frac{dx}{ds}\frac{dy}{dt}\right),$$

$$m\left(\frac{dx}{dt}\frac{d^2x}{dt^2} + \frac{dy}{dt}\frac{d^2y}{dt^2}\right) = m\frac{ds}{dt}\frac{d^2s}{dt^2} = m\left(X\frac{dx}{dt} + Y\frac{dy}{dt}\right) \dots (3),$$

or, as we may write it,

$$\frac{d^2s}{dt^2} = X\frac{dx}{ds} + Y\frac{dy}{ds},$$

which might at once have been obtained by taking the tangential acceleration.

Now, it will be shewn in Chap. IX, that if the forces resolved into X and Y are such as occur in nature,

$$Xdx + Ydy$$

is the complete differential of some function $\phi(x, y)$.

Integrating (3) on this hypothesis, we have

$$\left(\frac{dx}{dt}\right)^2 + \left(\frac{dy}{dt}\right)^2 = v^2 = 2\phi(x, y) + C \,\ldots\ldots\ldots\, (4),$$

supposing v to represent the entire velocity of the particle at time t.

Imagine the particle to start at the time $t = 0$, from a point whose co-ordinates are a, b, with a velocity V,

we have from (4),

$$V^2 = 2\phi(a, b) + C;$$

and therefore $v^2 = V^2 + 2\phi(x, y) - 2\phi(a, b) \ldots\ldots\, (5)$.

An interesting result, as it shews that a particle constrained to move under the action of the forces X, Y, along a particular path from the point a, b to the point x, y; has on arriving at the latter point, the square of its velocity increased by a quantity entirely independent of the path pursued.

127. *To find the pressure on the curve.*

Multiply equation (1) by $\frac{dy}{dt}$ and (2) by $\frac{dx}{dt}$, and subtract.

Then, noticing that

$$\frac{dy}{ds}\,\frac{dy}{dt} + \frac{dx}{ds}\,\frac{dx}{dt} = \frac{\left(\frac{ds}{dt}\right)^2}{\frac{ds}{dt}} = \frac{ds}{dt},$$

we have,

$$m\left(\frac{dy}{dt}\frac{d^2x}{dt^2}-\frac{dx}{dt}\frac{d^2y}{dt^2}\right)=m\left(X\frac{dy}{dt}-Y\frac{dx}{dt}\right)-R\frac{ds}{dt}.$$

But (TODHUNTER, *Diff. Calc.* Art. 322), if ρ be the radius of curvature of the constraining curve at the point x, y,

$$\rho=\frac{\left(\frac{ds}{dt}\right)^3}{\frac{dx}{dt}\frac{d^2y}{dt^2}-\frac{dy}{dt}\frac{d^2x}{dt^2}}.$$

Transforming by means of this, the above equation becomes

$$-\frac{m\left(\frac{ds}{dt}\right)^2}{\rho}=m\left(X\frac{dy}{ds}-Y\frac{dx}{ds}\right)-R;$$

$$\text{or, } R=m\,(X\sin\theta-Y\cos\theta)+\frac{mv^2}{\rho}\;\ldots\ldots\ldots\;(6).$$

The two parts of which this expression consists are evidently the resolved pressure on the curve produced by the forces X and Y, and the pressure due to the velocity only.

128. *To find the point where the particle will leave the constraining curve.*

For this it is evident that we have only to put $R=0$, as then the motion will be free.

This condition gives us

$$\frac{v^2}{\rho}=Y\cos\theta-X\sin\theta.$$

Now let F be the resultant of X and Y, then if Q be the chord of curvature at P parallel to F, Q is evidently

$$=2\rho\sin FPT'=2\rho\,.\sin\,(FPX-\theta)$$

$$=2\rho\,.\,\frac{Y\cos\theta-X\sin\theta}{\sqrt{(X^2+Y^2)}}.$$

Hence, $$v^2 = \frac{Q}{2}\sqrt{(X^2+Y^2)}$$

$$= 2 . \sqrt{(X^2+Y^2)} . \frac{Q}{4}$$

$$= 2 . F . \frac{Q}{4} .$$

Comparing this with the formula $v^2 = 2fs$ (Art. 36), we see that *the particle will leave the curve at a point where its velocity is such as would be produced by the resultant force then acting on it, if continued constant during its fall from rest through a space equal to* ¼ *of the chord of curvature parallel to that resultant.*

129. The formulæ just given are much simplified when we consider gravity as the only force acting. Taking in this case the axis of y vertical; our forces become

$$X = 0 \text{ and } Y = -g;$$

and the velocity, and pressure on the curve are given by

$$v^2 - V^2 = 2g(k-y), \text{ if } v = V \text{ when } y = k,$$

$$\text{and } R = mg \cos\theta + \frac{mv^2}{\rho}.$$

Suppose we change the origin to the point from which the particle's motion is supposed to commence; and take the axis of y vertically *downwards;* we shall evidently have

$$v^2 - V^2 = 2gy;$$

and if the particle start from rest

$$v^2 = 2gy.$$

This shews that the velocity depends merely on the distance beneath a horizontal plane through the original position of rest. Hence, whatever be the nature of the curve on which a particle slides under the action of gravity, its motion will always be in the same direction till it rises to the same level as that to the fall from which its velocity is due. If it cannot do so, its motion will be constantly in the same direction; if it can, its velocity will become zero, and the particle will *then* either come permanently to rest, or return to the point from which it started.

130. *To find the time of a particle's sliding down any arc of a curve from rest at the upper extremity.*

Taking the upper extremity as origin and the axis of y downwards; we have

$$\frac{ds}{dt} = v = \sqrt{(2gy)};$$

$$\text{and } \therefore t = \int_0^{y_1} \frac{\frac{ds}{dy}\,dy}{\sqrt{(2gy)}} \quad\text{.....................}\quad (1)$$

if y_1 define the other extremity of the arc of the given curve.

Or, taking the lower point as origin, and axis of y upwards, we have, since in this case v tends to decrease s,

$$t = \int_{y_1}^{0} \frac{-\frac{ds}{dy}\,dy}{\sqrt{\{2g\,(y_1 - y)\}}} = \int_0^{y_1} \frac{\frac{ds}{dy}\,dy}{\sqrt{\{2g\,(y_1 - y)\}}} \quad\text{......}\quad (2).$$

131. Ex. *To find the time of descending from rest at any point of an inverted cycloid to the vertex.*

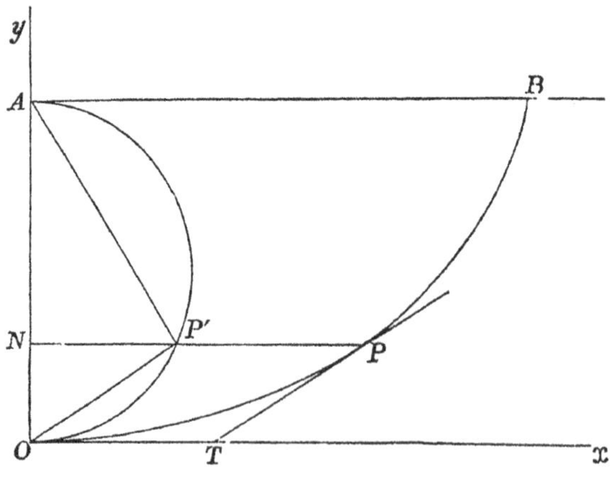

Taking formula (2); since in this case the vertex is origin, and axis that of y, we have from the figure

$$s = OP = 2 OP' = 2\sqrt{(AO \,.\, ON)} = 2\sqrt{(2ay)},$$

if a be the radius of the generating circle.

Hence, $$\frac{ds}{dy} = \sqrt{\left(\frac{2a}{y}\right)};$$

$$\text{and } t = +\sqrt{\left(\frac{a}{g}\right)} \int_0^{y_1} \frac{dy}{\sqrt{(yy_1 - y^2)}}$$

$$= \left\{C + \sqrt{\left(\frac{a}{g}\right)} \,.\, \text{vers}^{-1} \frac{2y}{y_1}\right\}_0^{y_1}$$

$$= \pi\sqrt{\left(\frac{a}{g}\right)},$$

which is independent of y_1 or of the point from which the particle begins its descent.

The reason of this remarkable property will be more easily seen if we take the formula for the acceleration in the direction of the arc. We have thus

$$\frac{d^2 s}{dt^2} = -g \sin (P'Ox)$$

(since OP' is parallel to the tangent to the cycloid at P)

$$= -g \,.\, \sin (OAP')$$

$$= -g \,.\, \frac{OP'}{OA}$$

$$= -g \,.\, \frac{s}{4a},$$

or the acceleration is proportional to the distance from the vertex. Comparing this with Arts. 42—44, the reason of the above result will be evident.

132. *A particle acted on by gravity moves in an arc of a vertical circle, to determine the motion.*

The equation to the circle from its lowest point is

$$x = \sqrt{(2ay - y^2)}.$$

$$\text{Hence } \frac{ds}{dy} = \frac{a}{\sqrt{(2ay - y^2)}}.$$

$$\text{But } \frac{ds}{dt} = -\sqrt{\{2g(y_1 - y)\}},$$

if we suppose the motion to commence at the point defined by y_1; and therefore

$$\frac{dt}{dy} = -\frac{a}{\sqrt{(2g)}} \cdot \frac{1}{\sqrt{\{(y_1 - y)(2ay - y^2)\}}} \quad \text{.......... (1).}$$

If we put $y = y_1 \sin^2\theta$, this becomes

$$t = -\sqrt{\left(\frac{a}{g}\right)} \int \frac{d\theta}{\sqrt{\left(1 - \frac{y_1}{2a}\sin^2\theta\right)}},$$

an elliptic integral of the first order, whose value for given limits can only be approximated to, except when $y_1 = 2a$, or the velocity is that due to a fall from the highest point of the circle, which case we will shortly consider (Art. 134).

(1) may be put in the form

$$\frac{dt}{dy} = -\frac{1}{2}\sqrt{\left(\frac{a}{g}\right)} \frac{1}{\sqrt{(y_1 y - y^2)}} \left(1 - \frac{y}{2a}\right)^{-\frac{1}{2}}$$

$$= -\frac{1}{2}\sqrt{\left(\frac{a}{g}\right)} \cdot \frac{1}{(y_1 y - y^2)^{\frac{1}{2}}} \left\{1 + \frac{1}{2} \cdot \frac{y}{2a} + \frac{1 \cdot 3}{2 \cdot 4}\left(\frac{y}{2a}\right)^2 \right.$$

$$\left. + \frac{1 \cdot 3 \cdot 5}{2 \cdot 4 \cdot 6}\left(\frac{y}{2a}\right)^3 + \ldots + \frac{1 \cdot 3 \ldots (2n-1)}{2 \cdot 4 \ldots 2n}\left(\frac{y}{2a}\right)^n + \&c.\right\},$$

each term of which may be integrated separately.

Suppose it be required to determine the time of descent to the lowest point, the limits of y are y_1 and 0. If we notice that

$$\int \frac{y^n dy}{\sqrt{(y_1 y - y^2)}} = \frac{2n-1}{2n} y_1 \int \frac{y^{n-1} dy}{\sqrt{(y_1 y - y^2)}} - \frac{y^{n-1}\sqrt{(y_1 y - y^2)}}{n};$$

$$\text{and } \therefore \int_{y_1}^{0} \frac{y^n dy}{\sqrt{(y_1 y - y^2)}} = \frac{2n-1}{2n} y_1 \int_{y_1}^{0} \frac{y^{n-1} dy}{\sqrt{(y_1 y - y^2)}},$$

$$\text{while } \int_{y_1}^{0} \frac{y dy}{\sqrt{(y_1 y - y^2)}} = \left(\frac{y_1}{2} \cdot \text{vers}^{-1} \frac{2y}{y_1} + C\right)_{y_1}^{0} = -\frac{\pi y_1}{2},$$

$$\text{we have } \int_{y_1}^{0} \frac{y^n dy}{\sqrt{(y_1 y - y^2)}} = -\frac{1 \cdot 3 \cdot 5 \ldots (2n-1)}{2 \cdot 4 \cdot 6 \ldots 2n} \pi y_1^n.$$

Hence time of fall to the lowest point

$$T = \frac{\pi}{2}\sqrt{\left(\frac{a}{g}\right)}\left[1 + \left(\frac{1}{2}\right)^2 \cdot \frac{y_1}{2a} + \left(\frac{1 \cdot 3}{2 \cdot 4}\right)^2 \cdot \left(\frac{y_1}{2a}\right)^2 + \ldots \right.$$
$$\left. + \left\{\frac{1 \cdot 3 \ldots (2n-1)}{2 \cdot 4 \ldots 2n}\right\}^2 \left(\frac{y_1}{2a}\right)^n + \ldots\right].$$

When the arc of vibration is very small we have

$$T = \frac{\pi}{2}\sqrt{\left(\frac{a}{g}\right)},$$

and the time of a complete oscillation is

$$2\pi\sqrt{\left(\frac{a}{g}\right)},$$

which coincides with that in a cycloid, observing that in the latter the quantity a is 4 times as great as here.

133. The next approximation gives a correction

$$\frac{\pi}{2}\sqrt{\left(\frac{a}{g}\right)} \cdot \frac{y_1}{8a},$$

whose ratio to the time of oscillation is

$$\frac{y_1}{8a} = (\tfrac{1}{4} \text{ chord semi-angle of oscillation})^2.$$

Thus if the particle oscillate through an arc whose chord is $\frac{1}{10}$, on each side of the vertical, the time of oscillation given by the formula $\pi\sqrt{\left(\frac{a}{g}\right)}$ will be incorrect by about $\frac{1}{1600}$ of its amount, in defect.

When the particle is supposed to be suspended by a thread without weight, it becomes what is termed a *simple pendulum*. Such a machine can exist only in theory, but Dynamics furnishes us with the means of reducing the calculation of the motion of such a pendulum as we can construct, to that of the simple pendulum. It is evident that by this means we may determine the value of g, if the length of the pendulum, its arc of oscillation, and the number of vibrations it makes in a given time, be known. Since gravity decreases (according to a known law) as we ascend above the Earth's surface, the comparison of the times of vibration of the same pendulum on the top of a mountain and at its base would give approximately the height.

134. Or we may take the equation for the acceleration along the arc.

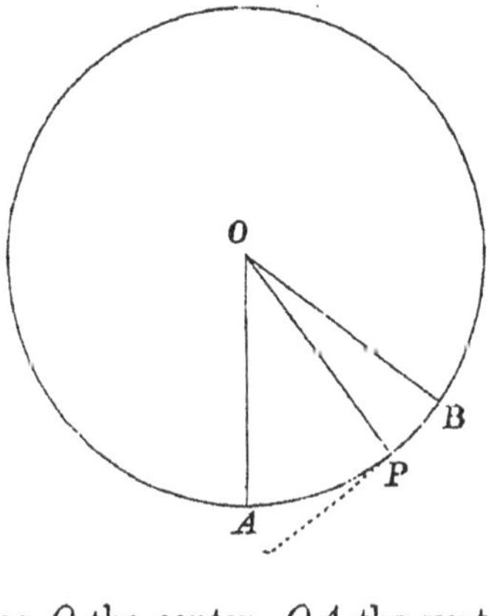

Thus suppose O the center, OA the vertical radius, B the point whence the particle starts with velocity $= a\omega$ at time $t = 0$, P its position at time t.

Let $\left.\begin{array}{l} AOB = \alpha, \\ AOP = \theta, \end{array}\right\} \; OA = a.$

Then $\dfrac{d^2s}{dt^2} = -g \sin\theta.$

But $s = a\theta$;

$$\therefore \frac{d^2\theta}{dt^2} = -\frac{g}{a}\sin\theta \text{ .... (1).}$$

Multiplying by $2\dfrac{d\theta}{dt}$ and integrating, we have

$$\overline{\frac{d\theta}{dt}}\Big|^2 = C + \frac{2g}{a}\cos\theta.$$

But $\dfrac{d\theta}{dt} = \omega$ when $\theta = \alpha$,

$$\therefore \frac{d\theta}{dt} = \sqrt{\left(\frac{2g}{a}\right)} \cdot \sqrt{\left(\cos\theta - \cos\alpha + \frac{a\omega^2}{2g}\right)} \text{...... (2).}$$

This cannot be integrated a second time unless

$$\frac{a\omega^2}{2g} - \cos\alpha = 1;$$

$$\text{or } a^2\omega^2 = 2g \cdot a(1 + \cos\alpha);$$

i.e. unless the velocity of projection at B, be that due to a fall through the difference of altitudes of B and the highest point of the circle.

In this case,

$$\frac{d\theta}{dt} = 2\sqrt{\left(\frac{g}{a}\right)} \cdot \cos\frac{\theta}{2}.$$

$$\text{And } \therefore \sqrt{\left(\frac{g}{a}\right)} \cdot t = \log_e C\sqrt{\frac{\left(1 + \sin\frac{\theta}{2}\right)}{\left(1 - \sin\frac{\theta}{2}\right)}}.$$

But $t=0$, $\theta=\alpha$, together,

$$\therefore \sqrt{\left(\frac{g}{a}\right)} . t = \log_e \sqrt{\frac{\left(1-\sin\frac{\alpha}{2}\right)\left(1+\sin\frac{\theta}{2}\right)}{\left(1+\sin\frac{\alpha}{2}\right)\left(1-\sin\frac{\theta}{2}\right)}} \ \ldots\ldots (3),$$

which determines the motion completely.

From the remark in (129), it is evident that after reaching A, the particle will ascend the other semicircle with a velocity just sufficient to carry it to the highest point; the time at which it will reach that point after leaving A, will be found by putting

$$\theta = \pi, \quad \alpha = 0, \quad \text{in } (3);$$

$$\text{or, } \sqrt{\left(\frac{g}{a}\right)} . T = \log_e \infty = \infty;$$

or the particle will continually approach the highest point, and never reach it.

135. *To find the pressure on the curve.*

Suppose R directed outwards from the center, then evidently

$$R = m\left(\frac{v^2}{a} + g\cos\theta\right)$$

$$= m\{2g(\cos\theta - \cos\alpha) + a\omega^2 + g\cos\theta\}, \text{ by } (2)$$

$$= m(3g\cos\theta - 2g\cos\alpha + a\omega^2).$$

Suppose the particle to have been projected from A, with velocity $a\omega$; then $\alpha = 0$;

$$\text{and } R = m(3g\cos\theta - 2g + a\omega^2).$$

This will have a value zero, if

$$a\omega^2 < 5g, \text{ or } a\omega < \sqrt{(5ga)}.$$

It may happen however that the points thus found may not lie within the arc which the particle passes over.

There are positions of rest (Art. 129) when $a\omega < 2\sqrt{(ga)}$. Now that the points where $R=0$, may lie within the limits of oscillation, the value of $\cos\theta$ for the former, must not be less than that for the latter;

$$\text{or,}\quad \frac{2g-a\omega^2}{3g} \not< \frac{2g-a\omega^2}{2g}.$$

This condition can only be satisfied by $2g-a\omega^2$ vanishing or becoming negative; or by

$$a\omega \not< \sqrt{(2ga)}.$$

Hence if the velocity of projection from the lowest point do not fall short of $\sqrt{(2ga)}$, and do not exceed $\sqrt{(5ga)}$, there will be a point in the path at which $R=0$; and if the particle be moving on the concave side of a smooth circle, or be attached by a string to a fixed point, the circular motion will cease at that point; the particle will fall off the circle in the one case, and the string will cease to be stretched in the other.

Beyond these limits it is evident that we shall have for velocity of projection $>\sqrt{(5ga)}$ continuous revolution in the circle, and for velocity of projection $<\sqrt{(2ga)}$ oscillations about the lowest point.

Also by what we have before shewn, if the particle be constrained by a circular tube, it will oscillate if the velocity at the lowest point is less than $2\sqrt{(ga)}$, if equal to $2\sqrt{(ga)}$ it will reach the highest point after the lapse of an infinite time, and if greater than $2\sqrt{(ga)}$ it will revolve continuously.

136. *A particle falls from rest at a height k down the semicubical parabola whose equation is $ax^2=y^3$, the axis of y being vertical; to determine the motion.*

Here, $\dfrac{ds}{dt}=-\sqrt{\{2g(k-y)\}}$, since gravity tends to diminish s.

$$\text{Also}\quad \frac{ds}{dy}=\sqrt{\left\{1+\left(\frac{dx}{dy}\right)^2\right\}}=\sqrt{\left(1+\frac{9y^4}{4a^2x^2}\right)}=\sqrt{\left(1+\frac{9y}{4a}\right)}.$$

Hence, $\frac{dy}{dt} = -\sqrt{\left\{\frac{8ga(k-y)}{4a+9y}\right\}}$;

and the time of fall to a point where $y = l$ is

$$t_1 = \frac{1}{\sqrt{(8ga)}} \int_l^k \sqrt{\left(\frac{4a+9y}{k-y}\right)} . dy.$$

Let $\frac{\theta^2}{9} = \frac{k-y}{4a+9y}$, the limits of θ are $\sqrt{\left\{\frac{9(k-l)}{4a+9l}\right\}}$ and 0;

$$\text{and therefore } t_1 = \frac{9k+4a}{3\sqrt{(8ga)}} \int_0^{\sqrt{\left\{\frac{9(k-l)}{4a+9l}\right\}}} . \frac{2d\theta}{(1+\theta^2)^2}.$$

Put $\theta = \tan\phi$, limits are $\tan^{-1}\sqrt{\left\{\frac{9(k-l)}{4a+9l}\right\}}$ and 0, and

$$t_1 = \frac{9k+4a}{3\sqrt{(8ga)}} \int_0^{\tan^{-1}\sqrt{\left\{\frac{9(k-l)}{4a+9l}\right\}}} 2d\phi \cos^2\phi$$

$$= \frac{9k+4a}{3\sqrt{(8ga)}} \left\{\tan^{-1}\sqrt{\left\{\frac{9(k-l)}{4a+9l}\right\}} + \frac{\sqrt{\left\{\frac{9(k-l)}{4a+9l}\right\}}}{1+\frac{9(k-l)}{4a+9l}}\right\};$$

which determines t_1 for any values of k and l.

If the time of fall to the cusp at the origin be required, $l = 0$, and

$$T = \frac{9k+4a}{3\sqrt{(8ga)}} \left\{\tan^{-1}\frac{3}{2}\sqrt{\left(\frac{k}{a}\right)} + \frac{6\sqrt{(ak)}}{4a+9k}\right\}.$$

137. *A particle acted on by gravity is projected from the vertex, along a parabola whose axis is vertical and vertex upwards; to determine the motion and the pressure on the curve.*

Let $x^2=4ay$ be the equation, the axis of y being vertically downwards, and the vertex the origin. Then

$$\left.\frac{ds}{dt}\right|^2 = v^2 = V^2 + 2gy,$$

where V is the velocity at the vertex. Suppose it due to a height l, then

$$\left.\frac{ds}{dt}\right|^2 = 2g\,(l+y) \quad\text{.....................}\quad (1).$$

$$\text{Now } \left(\frac{dy}{ds}\right)^2 = \frac{y}{a+y},$$

by the equation to the curve.

$$\text{Hence } \left.\frac{dy}{dt}\right|^2 = \frac{2gy\,(l+y)}{a+y},$$

or for the time of fall to a depth k

$$t_1 = \frac{1}{\sqrt{(2g)}} \int_0^k \sqrt{\left\{\frac{a+y}{y\,(l+y)}\right\}} \,.\, dy,$$

which is thus determined.

For the pressure on the curve, supposing it positive when *from* the axis,

$$R = m\left(\frac{v^2}{\rho} - g\,\frac{dx}{ds}\right)$$

$$= m\left\{\frac{2g\,(l+y)}{2a\left(\dfrac{a+y}{a}\right)^{\frac{3}{2}}} - g\left(\frac{a}{a+y}\right)^{\frac{1}{2}}\right\}$$

$$= mg\sqrt{\left(\frac{a}{a+y}\right)}\,\frac{l-a}{a+y}.$$

If $l>a$, R is positive, or the particle will move on the concave side of the curve; if $l<a$, R is negative, and the particle moves on the convex side. In each of these cases

the pressure is inversely as the $\frac{3}{2}$th power of the distance below the directrix.

If $l = a$, or the velocity of projection at the vertex be that due to a fall from the directrix, R is zero the whole way or the particle moves freely, as we might have inferred from the results of Chap. IV.

138. *To find a curve such that a particle under the action of gravity will descend any arc of it from a given point, in the same time as it takes to descend the chord of that arc.*

Take the vertical through the given point as the initial line, then if ρ, θ be the polar co-ordinates of a point in the curve, the given point being pole, the conditions of the problem give at once

$$\int_{\theta_0}^{\theta} \frac{\frac{ds}{d\theta}\, d\theta}{\sqrt{(2g\rho \cos \theta)}} = \sqrt{\left(\frac{2\rho}{g \cos \theta}\right)},$$

θ_0 being the inclination to the vertical of the tangent at the point of departure.

Or, differentiating with respect to θ,

$$\frac{\frac{ds}{d\theta}}{\sqrt{(2\rho \cos \theta)}} = \frac{\sqrt{(2)}}{2} \left\{ \frac{\frac{d\rho}{d\theta}}{\sqrt{(\rho \cos \theta)}} + \frac{\sqrt{(\rho)} \sin \theta}{(\cos \theta)^{\frac{3}{2}}} \right\}.$$

$$\text{Hence } \frac{ds}{d\theta} = \frac{d\rho}{d\theta} + \rho \tan \theta.$$

$$\text{But } \left.\frac{ds}{d\theta}\right|^2 = \rho^2 + \left.\frac{d\rho}{d\theta}\right|^2,$$

and therefore squaring both sides and reducing

$$2\frac{d\rho}{\rho} = 2\frac{\cos 2\theta \, d\theta}{\sin 2\theta},$$

$$\text{or } \log_e \rho^2 = \log_e C \sin 2\theta,$$

$$\text{that is, } \rho^2 = a^2 \sin 2\theta,$$

the Lemniscate of Bernoulli, the node being the pole, and a tangent at that point the initial line.

139. *To find a curve such that if a particle acted on by gravity, fall down it through a vertical space h, starting from the vertex with velocity due to a height h, the time of fall will be independent of h.*

Let the axis of y be vertically downwards, then evidently

$$t_1 = \int_0^h \frac{\frac{ds}{dy} dy}{\sqrt{\{2g(h+y)\}}}.$$

Let $\frac{ds}{dy} = \phi(y)$ be the required differential equation to the curve,

$$t_1 = \frac{1}{\sqrt{(2g)}} \int_0^h \frac{\phi(y)\, dy}{\sqrt{(h+y)}}$$

$$= \frac{1}{\sqrt{(2g)}} \int_0^1 \frac{\phi(hz)\, h^{\frac{1}{2}} . dz}{\sqrt{(1+z)}}, \quad \text{if } y = hz.$$

Now $\frac{dt_1}{dh}$ is to be identically zero, hence we have

$$hz\, \phi'(hz) = -\tfrac{1}{2} \phi(hz),$$

$$\text{or } \frac{\phi'(y)\, dy}{\phi(y)} = -\frac{dy}{2y},$$

$$\therefore \phi(y) = Cy^{-\frac{1}{2}} = \frac{ds}{dy},$$

which shews the curve to be a cycloid, whose base is horizontal and vertex upwards.

140. *Two points being given, which are neither in a vertical nor in a horizontal line, to find the curve joining them, down which a particle sliding under the action of gravity, and starting from rest at the highest, will reach the other in the least possible time.*

The curve must evidently lie in the vertical plane passing through the points. For suppose it not to lie in that plane, project it on the plane, and call corresponding elements of the curve and its projection σ and σ'. Then if a particle slide down the projected curve its velocity at σ' will be the same as the velocity in the other at σ. But σ is *never less* than σ', and is generally greater. Hence the time through σ' is generally less than that through σ, and *never greater*. That is, the whole time of falling through the projected curve is less than that through the curve itself. Or the required curve lies in the vertical plane through the points.

Taking the axes of x and y, horizontal and vertically downwards, from the starting point; if x_0 be the abscissa of the other point, the time of descent will be

$$\int_0^{x_0} \frac{\frac{ds}{dx}\,dx}{\sqrt{(2gy)}}, \quad \text{or writing } \frac{dy}{dx}=p,$$

$$\int_0^{x_0} \frac{\sqrt{(1+p^2)}}{\sqrt{(2gy)}}\,dx.$$

Applying the rules of the Calculus of Variations, we have since V or $\frac{\sqrt{(1+p^2)}}{\sqrt{(y)}}$ is a function of y and p, the condition for a minimum, (Airy's Tracts, p. 224) viz.

$$V=p\left(\frac{dV}{dp}\right)+C,$$

the differential coefficient being partial.

This gives $\dfrac{\sqrt{(1+p^2)}}{\sqrt{(y)}}=\dfrac{p^2}{\sqrt{(y)}\sqrt{(1+p^2)}}+C,$

$$\text{or } \sqrt{(y)}\,.\sqrt{(1+p^2)}=\frac{1}{C}=\sqrt{(a)} \text{ suppose.}$$

$$\text{Hence } \frac{ds}{dy}=\frac{\sqrt{(1+p^2)}}{p}=\sqrt{\left(\frac{a}{a-y}\right)},$$

the differential equation to a cycloid, the origin being a cusp and the base the axis of x.

This is a problem celebrated in the history of Dynamics. The cycloid has received on account of this property the name of Brachistochrone. Farther on we propose to investigate the nature and some of the properties of Brachistochrones for other forces besides gravity. For an investigation not involving the Calculus of Variations see Appendix.

141. *To find the curve down which if a particle projected with a given velocity slide under the action of gravity, it will descend equal vertical spaces in equal times.*

Here we have, taking the axis of x horizontal, and that of y vertically downwards,

$$\frac{ds}{dt} = \sqrt{(2gy)};$$

if the velocity is that due to a fall from the axis of x.

Also by condition $\frac{dy}{dt} = \text{const.} = \sqrt{(2gh)}$, suppose;

$$\therefore \frac{ds}{dy} = \sqrt{\left(\frac{y}{h}\right)},$$

$$\frac{dx}{dy} = \sqrt{\left(\frac{y-h}{h}\right)};$$

$$\text{or, } x + C = \frac{2}{3}\left\{\frac{(y-h)^3}{h}\right\}^{\frac{1}{2}};$$

the semicubical parabola.

If the horizontal velocity is to be constant, we have

$$\frac{dx}{dt} = \sqrt{(2gk)}; \quad \therefore \frac{ds}{dx} = \sqrt{\left(\frac{y}{k}\right)} \text{ or } \frac{dy}{dx} = \sqrt{\left(\frac{y-k}{k}\right)},$$

and therefore $2\sqrt{(k)}\sqrt{(y-k)} = x + C$;

a parabola with its axis vertical and vertex upwards; as indeed we might have foreseen from the results of Chap. IV.

142. PROP. *A particle moves on a smooth plane curve under the action of a force directed to a fixed center in the plane of the curve; to determine the motion.*

Let $r=f(\theta)$ be the polar equation to the constraining curve about the center of force as pole, and let $P=\phi(r)$ be the central *repulsive* force on a particle whose distance from the center is r.

Resolving along the tangent at any point,

$$\frac{d^2s}{dt^2}=P\frac{dr}{ds}\ \dots\dots\dots\dots\ (1).$$

Hence, $$\left.\frac{ds}{dt}\right|^2=v^2=C+2\int\phi(r)\,dr\ \dots\dots\dots\ (2).$$

Equation (2) contains the complete solution of the problem as far as the motion is concerned, since by means of the equation to the curve either r or s may be eliminated from it, and if the resulting differential equation is integrable, it will give s or r in terms of t.

For the pressure on the curve. Resolving along the normal at any point,

$$\frac{mv^2}{\rho}+mP\frac{rd\theta}{ds}=R\ \dots\dots\dots\ (3),$$

an expression which by means of the foregoing equations will give R in terms of t or r.

Hence the solution is complete.

143. *A particle initially at rest at a point of the logarithmic spiral $r=a\epsilon^{n\theta}$ whose radius vector is b, moves on the curve under the action of a center of force $\propto$ distance, situated at the pole; to determine the motion.*

Here $$\frac{d^2s}{dt^2}=-\mu r\frac{dr}{ds};$$

$$\therefore \left.\frac{ds}{dt}\right|^2 = C - \mu r^2,$$

$$\text{and } 0 = C - \mu b^2,$$

$$\frac{ds}{dt} = \sqrt{\{\mu (b^2 - r^2)\}}.$$

$$\text{But } \frac{ds}{dt} = \sqrt{\left\{\left.\frac{dr}{dt}\right|^2 + r^2 \left.\frac{d\theta}{dt}\right|^2\right\}}$$

$$= \frac{dr}{dt}\sqrt{\left(1 + \frac{1}{n^2}\right)};$$

$$\therefore \frac{dr}{\sqrt{(b^2 - r^2)}} = \frac{n\sqrt{(\mu)}}{\sqrt{(1+n^2)}}\,dt,$$

$$r = b \cos\left\{\frac{n\sqrt{(\mu)}\,t}{\sqrt{(1+n^2)}} + \beta\right\},$$

$$b = b\cos\beta; \quad \therefore \beta = 0,$$

$$\text{and } r = b \cos \frac{n\sqrt{(\mu)}\,t}{\sqrt{(1+n^2)}};$$

which gives the position of the particle at any time.

When it reaches the pole $r = 0$;

$$\therefore t' = \frac{\pi\sqrt{(1+n^2)}}{2n\sqrt{(\mu)}}.$$

For the pressure on the curve,

$$R = m\left(\frac{v^2}{\rho} - \mu r \frac{r d\theta}{ds}\right)$$

$$= m\left\{\frac{\mu (b^2 - r^2)}{r\sqrt{(1+n^2)}} - \frac{\mu r}{\sqrt{(1+n^2)}}\right\}$$

$$= \frac{m\mu}{\sqrt{(1+n^2)}}\,\frac{b^2 - 2r^2}{r}.$$

Hence the pressure is towards the pole when the motion commences, becomes zero when the particle's distance from the pole is diminished in the ratio of $\frac{1}{\sqrt{(2)}}$, and then is directed from the pole for the rest of the motion.

144. *When the constraining curve is one of double curvature.*

All we know directly about R is that it is perpendicular to the tangent line at any point.

Resolve then the given forces acting upon the particle into three, one along the tangent equal to S which in all cases in nature will be a function of x, y, z and therefore of s, another in the line of intersection of the normal and osculating planes (or radius of absolute curvature) equal to T, and the third equal to P perpendicular to each of the other two.

Let the resolved parts of R in the directions of T and P be R_1, R_2. Then the acceleration in the first direction is $\frac{d^2s}{dt^2}$, and therefore

$$\frac{d^2s}{dt^2} = S. \qquad (1).$$

This equation together with the two to the curve is sufficient to determine the *motion* completely.

Now the particle at any point of its path may be supposed to move in the osculating plane. Hence, by our investigation for the motion on a plane curve, if ρ be the radius of absolute curvature, v the velocity,

$$R_1 = \frac{mv^2}{\rho} - mT \qquad (2),$$

T being considered positive when it acts towards the center of absolute curvature.

Now R_2 is the force which prevents P's withdrawing the particle from the osculating plane;

$$\therefore R_2 = -mP \qquad (3),$$

(2) and (3) give the resolved parts of the pressure on the curve.

Also $R = \sqrt{(R_1^2 + R_2^2)}$, and makes an angle $= \tan^{-1}\left(\frac{R_2}{R_1}\right)$ with the osculating plane.

145. In Art. 131 we arrived at the remarkable property of the cycloid, that a particle falling under the action of gravity from rest at any point of the curve reaches the lowest point in the same time. Let us find for what forces the same property is possessed by any other given curve.

Let the accelerating forces resolved along the curve have a resultant $= \phi'(s)$, where s is the distance at any instant from the lowest point, then

$$\frac{d^2s}{dt^2} = S = -\phi'(s) \qquad (1);$$

and if the particle starts at a distance k from the lowest point, the velocity $= 0$ when $s = k$. Hence the corrected integral of (1) is

$$\left(\frac{ds}{dt}\right)^2 = 2\{\phi(k) - \phi(s)\},$$

$$\text{or } t_1 = \int_k^0 \frac{-ds}{\sqrt{(2)}.\{\phi(k) - \phi(s)\}^{\frac{1}{2}}};$$

which is by hypothesis to be independent of k.

Put $s = kx$, the limits of x are 1 and 0, and

$$t_1 \propto \int_0^1 \frac{kdx}{\{\phi(k) - \phi(kx)\}^{\frac{1}{2}}};$$

$$\text{and } \therefore \frac{dt_1}{dk} \propto \int_0^1 \frac{\{\phi(k) - \phi(kx)\}^{\frac{1}{2}} - \frac{k}{2}.\frac{\phi'(k) - \phi'(kx)\,x}{\{\phi(k) - \phi(kx)\}^{\frac{1}{2}}}}{\phi(k) - \phi(kx)}\,dx$$

$= 0$ identically.

$$\text{Hence } \left\{\phi(k) - \frac{k}{2}.\phi'(k)\right\} - \left\{\phi(kx) - \frac{kx}{2}.\phi'(kx)\right\} = 0$$

identically, which can only be the case if

$$\phi(k) = \frac{k}{2}\phi'(k),$$

or if $\phi(k) =$ constant, which case we evidently need not consider.

$$\text{Hence } \frac{\phi'(k)}{\phi(k)} = \frac{2}{k},$$

$$\text{or } \log_e \phi(k) = \log_e C'k^2,$$

$$\text{or } \phi(k) = C'k^2,$$

$$\text{and } \phi'(k) = Ck.$$

$$\text{Hence } \frac{d^2 s}{dt^2} = S = -Cs \ldots\ldots\ldots\ldots\ldots\ldots (2),$$

or the resolved force along the curve must be proportional to the distance from the point of departure.

146. We might have arrived at the same conclusion, but not quite so satisfactorily, thus,

$$t_1 = \frac{1}{\sqrt{(2)}} \int_0^k \frac{ds}{\{\phi(k)\}^{\frac{1}{2}}} \left[1 + \tfrac{1}{2}\frac{\phi(s)}{\phi(k)} + \ldots + N\left\{\frac{\phi(s)}{\phi(k)}\right\}^n + \ldots\right].$$

Now the condition that the $(n+1)^{\text{th}}$ term when integrated between the limits should not contain k is

$$\int_0^k \frac{\{\phi(s)\}^n ds}{\{\phi(k)\}^{n+\frac{1}{2}}} \text{ independent of } k.$$

This can only be the case if $\{\phi(k)\}^{\frac{1}{2}}$ and $\therefore$ $\{\phi(s)\}^{\frac{1}{2}}$ be of the same dimensions as ds, and therefore as s.

$$\text{Hence take } \{\phi(s)\}^{\frac{1}{2}} \propto s,$$

$$\text{or } \phi(s) = Cs^2;$$

$$\therefore \phi'(s) = C's \text{ as before.}$$

Hence if X, Y, Z be the impressed forces,

$$X\frac{dx}{ds}+Y\frac{dy}{ds}+Z\frac{dz}{ds}=-Cs$$

is the condition they must satisfy at every point xyz of the given curve. For such forces the given curve is said to be a Tautochrone.

By equation (2) Art. 145, the time of descent is

$$\tau=\frac{\pi}{2\sqrt{(C)}}.\quad \text{Hence } C=\frac{\pi^2}{4\tau^2}.$$

147. *To find the Brachistochrone for a particle subjected to the action of any forces which make $Xdx+Ydy+Zdz$ a perfect differential.*

Generally $t=\int\frac{ds}{v}$ between proper limits, is to be a minimum;

$$\therefore\ \delta t=\int\frac{v\delta ds-ds\delta v}{v^2}=0\ \ldots\ldots\ldots\ldots\ldots\ldots\ (1).$$

But here $\quad v^2=2\int(Xdx+Ydy+Zdz)$;

$$\therefore\ v\delta v=X\delta x+Y\delta y+Z\delta z,$$

$$\text{or } ds\delta v=(X\delta x+Y\delta y+Z\delta z)\,dt\ \ldots\ldots\ldots\ldots\ (2).$$

$$\text{Again } ds^2=dx^2+dy^2+dz^2,$$

$$\text{and } \frac{ds}{dt}\delta ds=v\delta ds=\frac{dx}{dt}\delta dx+\frac{dy}{dt}\delta dy+\frac{dz}{dt}\delta dz\ \ldots\ldots\ (3).$$

Hence (1) becomes, by (2) and (3), and since the order of d and δ is immaterial,

$$0=\int\frac{1}{v^2}\left(\frac{dx}{dt}d\delta x+\frac{dy}{dt}d\delta y+\frac{dz}{dt}d\delta z\right)$$
$$-\int\frac{1}{v^2}(X\delta x+Y\delta y+Z\delta z)\,dt$$

$$=\frac{1}{v^2}\left(\frac{dx}{dt}\delta x+\frac{dy}{dt}\delta y+\frac{dz}{dt}\delta z\right)$$

$$-\int dt\left[\left\{\frac{d}{dt}\left(\frac{1}{v^2}\frac{dx}{dt}\right)+\frac{X}{v^2}\right\}\delta x+\left\{\frac{d}{dt}\left(\frac{1}{v^2}\frac{dy}{dt}\right)+\frac{Y}{v^2}\right\}\delta y\right.$$
$$\left.+\left\{\frac{d}{dt}\left(\frac{1}{v^2}\frac{dz}{dt}\right)+\frac{Z}{v^2}\right\}\delta z\right],$$

by integrating the first term by parts.

But evidently the integrated part vanishes at both limits, and in order that the other may be identically zero, we must have

$$\frac{d}{dt}\left(\frac{1}{v^2}\frac{dx}{dt}\right)+\frac{X}{v^2}=0 \text{ .... } (4),$$

with similar expressions in y and z. The elimination of t and v or $\frac{ds}{dt}$ from these equations will give us two differential equations to the curve required.

But without getting rid of v we may prove two properties common to all such Brachistochrones.

Eliminating t from (4) we have

$$v\frac{d}{ds}\left(\frac{1}{v}\frac{dx}{ds}\right)+\frac{X}{v^2}=0,$$

$$\text{or } v^2\frac{d^2x}{ds^2}-v\frac{dv}{ds}\frac{dx}{ds}+X=0 \text{ .... } (5),$$

with similar expressions in y and z.

Multiplying these in order by λ, μ, ν and adding; if we take λ, μ, ν such that

$$\left.\begin{aligned}\lambda\frac{d^2x}{ds^2}+\mu\frac{d^2y}{ds^2}+\nu\frac{d^2z}{ds^2}=0\\ \lambda\frac{dx}{ds}+\mu\frac{dy}{ds}+\nu\frac{dz}{ds}=0\end{aligned}\right\} \text{ .... } (6),$$

we shall have also

$$\lambda X + \mu Y + \nu Z = 0 \text{ .... } (7).$$

Now (6) shews that the line whose direction cosines are as λ, μ, ν is perpendicular to the radius of absolute curvature of the path, and also to the tangent; that is, it is normal to the osculating plane. Also by (7) the same line is perpendicular to the resultant of X, Y, Z.

Hence, *the osculating plane at any point contains the resultant of the impressed forces.*

Now if ρ be the radius of absolute curvature

$$\rho = \left\{\left(\frac{d^2x}{ds^2}\right)^2 + \left(\frac{d^2y}{ds^2}\right)^2 + \left(\frac{d^2z}{ds^2}\right)^2\right\}^{-\frac{1}{2}},$$

and its direction cosines as before are

$$\rho\frac{d^2x}{ds^2}, \quad \rho\frac{d^2y}{ds^2}, \quad \rho\frac{d^2z}{ds^2};$$

therefore multiplying equations (5) by

$$\frac{d^2x}{ds^2}, \quad \frac{d^2y}{ds^2}, \quad \frac{d^2z}{ds^2},$$

and adding, noting that

$$\frac{dx}{ds}\frac{d^2x}{ds^2} + \frac{dy}{ds}\frac{d^2y}{ds^2} + \frac{dz}{ds}\frac{d^2z}{ds^2} = 0,$$

we have

$$\frac{v^2}{\rho} = -\left(X\rho\frac{d^2x}{ds^2} + Y\rho\frac{d^2y}{ds^2} + Z\rho\frac{d^2z}{ds^2}\right),$$

or, *the portion of the pressure due to the velocity is equal to that produced by the impressed forces.*

148. *A particle moves under the action of given forces on a given smooth surface; to determine the motion.*

Let $F(x, y, z) = 0$, (1) be the equation to the surface, R the force acting in the normal to the surface, which is the only effect of the constraint. Then if λ, μ, ν be its direction cosines, we know that

$$\lambda = \frac{\left(\frac{dF}{dx}\right)}{\sqrt{\left\{\left(\frac{dF}{dx}\right)^2 + \left(\frac{dF}{dy}\right)^2 + \left(\frac{dF}{dz}\right)^2\right\}}} \quad \text{.........} \quad (2),$$

with similar expressions for μ and ν; the differential coefficients being partial.

X, Y, Z being as before the impressed forces, our equations of motion are evidently,

$$\left.\begin{aligned} m\frac{d^2x}{dt^2} &= mX + R\lambda \\ m\frac{d^2y}{dt^2} &= mY + R\mu \\ m\frac{d^2z}{dt^2} &= mZ + R\nu \end{aligned}\right\} \quad \text{...............} \quad (3).$$

Multiplying equations (3) respectively by

$$2\frac{dx}{dt}, \; 2\frac{dy}{dt}, \; 2\frac{dz}{dt},$$

and adding, we obtain

$$2m\left(\frac{dx}{dt}\frac{d^2x}{dt^2} + \frac{dy}{dt}\frac{d^2y}{dt^2} + \frac{dz}{dt}\frac{d^2z}{dt^2}\right)$$

$$= m\frac{d(v^2)}{dt} = 2m\left(X\frac{dx}{dt} + Y\frac{dy}{dt} + Z\frac{dz}{dt}\right) \quad \text{..........} \quad (4),$$

since the coefficient of R, or

$$\lambda\frac{dx}{dt} + \mu\frac{dy}{dt} + \nu\frac{dz}{dt},$$

vanishes because the line whose direction cosines are proportional to $\frac{dx}{dt}$, &c. being the tangent to the path, is perpendicular to the normal to the surface.

If we suppose X, Y, Z to be such forces as occur in nature, (Chap. IX.) the integral of (4) will be of the form,

$$\left.\frac{ds}{dt}\right|^2 = v^2 = 2\phi(x, y, z) + C \ldots\ldots\ldots\ldots\ldots\ (5),$$

and the velocity at any point will depend only on the initial circumstances of projection, and not on the form of the path pursued.

To find R, multiply equations (3) in order by λ, μ, ν, add and observe that $\lambda^2 + \mu^2 + \nu^2 = 1$. We thus obtain

$$m\left(\lambda\frac{d^2x}{dt^2} + \mu\frac{d^2y}{dt^2} + \nu\frac{d^2z}{dt^2}\right) = m(X\lambda + Y\mu + Z\nu) + R.$$

But if the radius of curvature of the normal section of $F(x, y, z) = 0$, through the element δs be ρ, we have

$$\rho = \frac{1}{\lambda\frac{d^2x}{ds^2} + \mu\frac{d^2y}{ds^2} + \nu\frac{d^2z}{ds^2}}.$$

Hence, since $\frac{d^2x}{dt^2} = \frac{d^2x}{ds^2}\left(\frac{ds}{dt}\right)^2 + \frac{dx}{ds}\frac{d^2s}{dt^2}$, &c.

and $\lambda\frac{dx}{ds} + \mu\frac{dy}{ds} + \nu\frac{dz}{ds} = 0$,

$$\frac{mv^2}{\rho} = (X\lambda + Y\mu + Z\nu) + R,$$

which gives the normal pressure on the surface.

149. *To find the curve which the particle describes on the surface.*

For this we must eliminate R from equations (3).

The result is

$$\frac{\frac{d^2x}{dt^2} - X}{\lambda} = \frac{\frac{d^2y}{dt^2} - Y}{\mu} = \frac{\frac{d^2z}{dt^2} - Z}{\nu} \ldots\ldots\ldots\ (6),$$

two equations, between which if t be eliminated, the result is the differential equation to a second surface intersecting the first in the curve described.

150. So far for the general case, let us now make particular hypotheses.

If there be no impressed forces on the particle, we have by (5), $v^2 = \left.\frac{ds}{dt}\right]^2 = C$, and equations (6) become

$$\frac{\frac{d^2x}{ds^2}}{\lambda} = \frac{\frac{d^2y}{ds^2}}{\mu} = \frac{\frac{d^2z}{ds^2}}{\nu}.$$

Now $\frac{d^2x}{ds^2}$, &c. are proportional to the direction cosines of the radius of absolute curvature of the path; λ, μ, ν are those of the normal to the surface. Hence those lines coincide, or the normal to the surface lies in the osculating plane to the path.

But this is the property of the longest or shortest line joining two points on a surface, hence we have the following.

If a particle subject to no forces move from one point to another of a smooth surface the length of the path described will be a maximum or minimum.

This result will be afterwards deduced from a different principle.

151. *A particle moves on a surface of revolution, the only force acting being gravity parallel to the axis of the surface; to determine the motion.*

Take the axis of the surface as that of z, the equation will be $F(x, y, z) = f\{\sqrt{(x^2 + y^2)}\} - z = 0$.

This may be written $f(\rho) - z = 0$, if ρ be the distance of any point in the surface from the axis.

Equations (6) become

$$\frac{\frac{d^2x}{dt^2}}{f'(\rho)\frac{x}{\rho}} = \frac{\frac{d^2y}{dt^2}}{f'(\rho)\frac{y}{\rho}} = \frac{\frac{d^2z}{dt^2} - g}{-1} \text{............} (7),$$

whence we have

$$x\frac{d^2y}{dt^2} - y\frac{d^2x}{dt^2} = 0.$$

But if θ be the angle between the plane containing ρ and the axis of z, and a fixed plane; we see (Art. 85) that this is equivalent to

$$\rho^2 . \frac{d\theta}{dt} = \text{const.} = h \text{..................} (8).$$

Now $\frac{dz}{dt} = f'(\rho) . \frac{d\rho}{dt} = f'(\rho) . \frac{d\rho}{d\theta} \frac{d\theta}{dt} = \frac{hf'(\rho)}{\rho^2} . \frac{d\rho}{d\theta}$.

And therefore $\frac{d^2z}{dt^2} = \frac{h^2}{\rho^2} . \frac{d}{d\theta}\left\{\frac{f'(\rho)}{\rho^2}\frac{d\rho}{d\theta}\right\}$.

But by equation (7), multiplying numerator and denominator of each of the first members by x and y respectively, and taking half the sum,

$$\frac{x\frac{d^2x}{dt^2} + y\frac{d^2y}{dt^2}}{2\rho f'(\rho)} = g - \frac{d^2z}{dt^2} \text{............} (9).$$

Now $x^2 + y^2 = \rho^2$,

$$x\frac{dx}{dt} + y\frac{dy}{dt} = \rho\frac{d\rho}{d\theta} . \frac{d\theta}{dt} = \frac{h}{\rho}\frac{d\rho}{d\theta};$$

$$\therefore\ x\frac{d^2x}{dt^2} + y\frac{d^2y}{dt^2} + \left.\overline{\frac{dx}{dt}}\right|^2 + \left.\overline{\frac{dy}{dt}}\right|^2 = \frac{h^2}{\rho^2}\frac{d}{d\theta}\left(\frac{1}{\rho}\frac{d\rho}{d\theta}\right),$$

and (9) becomes

$$\frac{\frac{h^2}{\rho^2}\cdot\frac{d}{d\theta}\left(\frac{1}{\rho}\frac{d\rho}{d\theta}\right)-\frac{h^2}{\rho^4}\left(\rho^2+\overline{\frac{d\rho}{d\theta}}\Big|^2\right)}{2\rho\,.f'(\rho)}=g-\frac{h^2}{\rho^2}\frac{d}{d\theta}\left\{\frac{f'(\rho)}{\rho^2}\frac{d\rho}{d\theta}\right\};$$

$$\text{or,}\quad \rho\frac{d^2\rho}{d\theta^2}-2\left(\frac{d\rho}{d\theta}\right)^2-\rho^2=\frac{2\rho^5f'(\rho)}{h^2}\left[g-\frac{h^2}{\rho^2}\frac{d}{d\theta}\left\{\frac{f'(\rho)}{\rho^2}\frac{d\rho}{d\theta}\right\}\right],$$

the differential equation to the projection of the path on the plane of xy.

152. *Suppose the motion to take place in a spherical bowl.*

Let us take the center as origin, and the axis of z vertically downwards.

Then $F(x, y, z) = x^2 + y^2 + z^2 - a^2 = 0$,

and the equations of motion are

$$m\,.\frac{d^2x}{dt^2}=-R\frac{x}{a},$$

$$m\,.\frac{d^2y}{dt^2}=-R\frac{y}{a},$$

$$m\,.\frac{d^2z}{dt^2}=g-R\frac{z}{a}.$$

$$\text{Hence,}\quad \overline{\frac{dx}{dt}}\Big|^2+\overline{\frac{dy}{dt}}\Big|^2+\overline{\frac{dz}{dt}}\Big|^2=C+2gz$$

$$=V^2-2g(k-z)\;\ldots\ldots\;(1),$$

if V and k are the initial values of v and z.

$$\text{But}\quad x\frac{d^2y}{dt^2}-y\frac{d^2x}{dt^2}=0;$$

$$\text{or,}\quad x\frac{dy}{dt}-y\frac{dx}{dt}=h.\;\ldots\ldots\ldots\ldots\ldots\;(2).$$

$$\text{Also } x\frac{dx}{dt} + y\frac{dy}{dt} + z\frac{dz}{dt} = 0,$$

by the equation to the surface.

Hence, eliminating $\frac{dx}{dt}$ and $\frac{dy}{dt}$, we have

$$t = \int \frac{a\,dz}{\sqrt{[(a^2 - z^2)\{V^2 - 2g(k-z)\} - h^2]}} \cdots\cdots (3),$$

an elliptic function which if it were integrable would give z, and consequently x and y in terms of t.

153. If the oscillations about the lowest point be very small, we may obtain interesting results by an approximate solution.

Let θ be the angle made by the radius through the particle with the axis of z, ψ the angle denoting the position of the plane containing these two lines, ρ the distance of the particle from the axis. Let the projection be made horizontally with velocity V when $\theta = \alpha$, $\psi = 0$, $t = 0$ together.

$$\text{Now } z = a\cos\theta = a\left(1 - \frac{\theta^2}{2}\right), \text{ approximately,}$$

$$k = a\cos\alpha = a\left(1 - \frac{\alpha^2}{2}\right) \cdots\cdots\cdots\cdots\cdots\cdots$$

Also (2) gives at once,

$$(4)\quad \rho^2\frac{d\psi}{dt} = h = a^2\sin^2\alpha \cdot \frac{V}{a\sin\alpha} = aV\alpha, \text{ approximately.}$$

$$\text{Hence by (3), } t = \int \frac{a^2\theta\,d\theta}{\sqrt{[a^2\theta^2\{V^2 - ga(\theta^2 - \alpha^2)\} - a^2V^2\alpha^2]}};$$

$$\text{or (putting } \frac{V^2}{ga} = \beta^2) = \sqrt{\left(\frac{a}{g}\right)} \int \frac{\theta\,d\theta}{\sqrt{\{(\alpha^2 - \theta^2)(\theta^2 - \beta^2)\}}} \cdots (5).$$

Hence, θ is never greater than α nor less than β.

If therefore $\alpha=\beta$, or $V^2=ga\alpha^2$, which is that due to the vertical height above the lowest point of the bowl, the particle will move in a *horizontal* circle, and therefore with uniform velocity. This is called a conical pendulum, as the constraint may be produced by a string attached to the center, and to the particle as well as by the sphere.

We may now put (5) into the form

$$t=\sqrt{\left(\frac{a}{g}\right)}\int\frac{\theta d\theta}{\sqrt{\left\{\left(\frac{\alpha^2-\beta^2}{2}\right)^2-\left(\theta^2-\frac{\alpha^2+\beta^2}{2}\right)^2\right\}}};$$

$$\text{and if } \theta^2=\frac{(\alpha^2+\beta^2)+\omega(\alpha^2-\beta^2)}{2},$$

we have

$$2t=\sqrt{\left(\frac{a}{g}\right)}\int\frac{d\omega}{\sqrt{(1-\omega^2)}};$$

$$\therefore\ 2(t+C)=\sqrt{\left(\frac{a}{g}\right)}\cos^{-1}\omega.$$

$$\text{But } t=0,\quad \theta=\alpha,\quad \omega=1;$$

$$\therefore\ \omega=\cos 2\sqrt{\left(\frac{g}{a}\right)}\,.\,t;$$

$$\text{or, } \theta^2=\frac{\alpha^2+\beta^2}{2}+\frac{\alpha^2-\beta^2}{2}\cos 2\sqrt{\left(\frac{g}{a}\right)}\,.\,t;$$

or, substituting for the cosine of the double arc,

$$\theta^2=\alpha^2\cos^2\sqrt{\left(\frac{g}{a}\right)}\,.\,t+\beta^2\sin^2\sqrt{\left(\frac{g}{a}\right)}\,.\,t\ \ldots\ldots\ldots\ (6).$$

The value of θ^2 is therefore periodic, for $t=0$, $\pi\sqrt{\left(\frac{a}{g}\right)}$, $2\pi\sqrt{\left(\frac{a}{g}\right)}$, &c. we have $\theta=\alpha$; and for $t=\frac{\pi}{2}\sqrt{\left(\frac{a}{g}\right)}$, $\frac{3\pi}{2}\sqrt{\left(\frac{a}{g}\right)}$, &c. $\theta=\beta$. Hence the period is $\pi\sqrt{\left(\frac{a}{g}\right)}$.

154. To find the motion of the plane in which θ is measured, we return to the equation (4),

$$\rho^2 \cdot \frac{d\psi}{dt} = aV\alpha; \quad \therefore \ d\psi = \frac{V\alpha dt}{a\theta^2} = \sqrt{\left(\frac{g}{a}\right)} \cdot \frac{\alpha\beta}{\theta^2} \cdot dt,$$

or by (6)

$$= \sqrt{\left(\frac{g}{a}\right)} \cdot \alpha\beta \frac{dt}{\alpha^2 \cos^2 \sqrt{\left(\frac{g}{a}\right)} \cdot t + \beta^2 \sin^2 \sqrt{\left(\frac{g}{a}\right)} \cdot t};$$

the integral of which (Art. 101) is

$$\psi + C = \tan^{-1}\left\{\frac{\beta}{\alpha} \tan \sqrt{\left(\frac{g}{a}\right)} \cdot t\right\}.$$

But $\psi = 0$, $t = 0$ together;

$$\therefore \ \tan \psi = \frac{\beta}{\alpha} \tan \sqrt{\left(\frac{g}{a}\right)} \cdot t. \ \ldots\ldots\ldots\ldots \ (7).$$

It is easy from this to deduce the following results, viz. that each quarter revolution of this plane is accomplished in the same time, and simultaneously with the change of θ in that plane from α to β, or β to α. Also that whatever we take as the initial position, the time of this plane's turning through two right angles is the same, namely, $\pi \sqrt{\left(\frac{a}{g}\right)}$.

155. If we eliminate t between (6) and (7), we find

$$\theta^2 = \frac{\alpha^2\beta^2}{\alpha^2 \sin^2 \psi + \beta^2 \cos^2 \psi}.$$

This is of the same form as the polar equation to an ellipse about the center. The projection of the particle's path on a horizontal plane is therefore approximately an ellipse, its semiaxes being $a\alpha$, $a\beta$.

156. Prop. *To determine approximately the apsidal angle.*

At an apse z is of course a maximum or minimum, and therefore $\frac{dz}{dt} = 0$. This gives

$$(a^2 - z^2)\{V^2 - 2g(k - z)\} - h^2 = 0 \ \ldots\ldots\ldots\ldots \ (1),$$

whose two positive roots α, β are the alternate values of z at the apses. Suppose the particle projected horizontally, the point of projection will be an apse; and if α be the root whose value is k,

$$h^2 = V^2 (a^2 - \alpha^2);$$

therefore (1) becomes after reduction

$$(\alpha - z) \{(\alpha + z) V^2 - 2g (a^2 - z^2)\} = 0 \ldots\ldots\ldots (2).$$

$$\text{Hence } V^2 = \frac{2g (a^2 - \beta^2)}{\alpha + \beta},$$

$$\text{and } h^2 = \frac{2g (a^2 - \alpha^2)(a^2 - \beta^2)}{\alpha + \beta}.$$

Also if $-\gamma$ be the third root of (1),

$$\gamma - \beta = \frac{V^2}{2g}, \text{ or by (2)}, = \frac{a^2 - \beta^2}{\alpha + \beta}.$$

$$\text{Hence } a + \gamma = \frac{(a + \alpha)(a + \beta)}{\alpha + \beta}.$$

$$\text{Now } \frac{d\psi}{dt} = \frac{h}{a^2 - z^2};$$

$$\therefore \frac{d\psi}{dz} = \frac{h}{a^2 - z^2} \cdot \frac{a}{\sqrt{\{2g (z - \alpha)(\beta - z)(\gamma + z)\}}}.$$

Hence the apsidal angle or the value of ψ from $z = \alpha$ to $z = \beta$, is

$$\psi_0 = a \sqrt{\left\{\frac{(a^2 - \alpha^2)(a^2 - \beta^2)}{\alpha + \beta}\right\}} \int_\alpha^\beta \frac{dz}{(a^2 - z^2)\sqrt{\{(z - \alpha)(\beta - z)(\gamma + z)\}}}.$$

To get rid of γ put $z = a - \omega$, the integral becomes

$$\int_{a-\beta}^{a-\alpha} \frac{d\omega}{\omega(2a - \omega)\sqrt{\left[\{(a - \alpha) - \omega\}\{\omega - (a - \beta)\}\left\{\frac{(a + \alpha)(a + \beta)}{\alpha + \beta} - \omega\right\}\right]}};$$

and, expanding in powers of ω those terms whose variation is small compared with themselves, we have finally

$$\psi_0 = \tfrac{1}{2}\sqrt{\{(a-\alpha)(a-\beta)\}}\int_{a-\beta}^{a-\alpha}\frac{d\omega}{\omega\sqrt{[\{(a-\alpha)-\omega\}\{\omega-(a-\beta)\}]}}$$

$$\left[1+\frac{\omega}{2}\left\{\frac{1}{a}+\frac{\alpha+\beta}{(a+\alpha)(a+\beta)}\right\}+\&c.\right]$$

$$=\frac{\pi}{2}\left[1+\tfrac{1}{2}\sqrt{\{(a-\alpha)(a-\beta)\}}\left\{\frac{1}{a}+\frac{\alpha+\beta}{(a+\alpha)(a+\beta)}\right\}+\ldots\right].$$

The integration may easily be carried on farther, all the terms being evidently positive, but we have enough to shew that the apsidal angle is greater than $\frac{\pi}{2}$, and that therefore in the approximate elliptic path considered in last article the apse continually progredes.

In the case of this orbit, if p and q be its semiaxes, we have

$$\left.\begin{aligned} a-\alpha &= \frac{p^2}{2a} \text{ nearly} \\ a-\beta &= \frac{q^2}{2a} \quad \ldots\ldots \end{aligned}\right\};$$

and therefore the apsidal angle

$$=\frac{\pi}{2}\left(1+\frac{3}{8}\cdot\frac{pq}{a^2}+\ldots\right),$$

or the rate of progression of the apse therefore varies as the area of the projected orbit nearly.

157. *To find the Brachistochrone for a particle constrained to move on a given smooth surface, gravity being the only impressed force.*

Let $F=0$ (1) be the equation to the given surface, z being the vertical axis.

Then $\frac{ds}{dt} = \sqrt{\{2g(z - z_0)\}}$,

and therefore the time between the given points is

$$t_0 = \int_{z_0}^{z_1} \frac{\frac{ds}{dz}\,dz}{\sqrt{\{2g(z - z_0)\}}} \quad \ldots\ldots\ldots\ldots\ldots\ldots (2).$$

For the condition that (2) be a minimum we obtain

$$\frac{d}{dz}\left\{\frac{\frac{dx}{ds}}{\sqrt{(z - z_0)}}\right\} . \delta x + \frac{d}{dz}\left\{\frac{\frac{dy}{ds}}{\sqrt{(z - z_0)}}\right\} . \delta y = 0.$$

But δx and δy are not independent, (1) gives us

$$\left(\frac{dF}{dx}\right) \delta x + \left(\frac{dF}{dy}\right) \delta y = 0.$$

Hence, eliminating, we obtain

$$\frac{\frac{d}{dz}\left\{\frac{dx}{ds}\frac{1}{\sqrt{(z - z_0)}}\right\}}{\left(\frac{dF}{dx}\right)} = \frac{\frac{d}{dz}\left\{\frac{dy}{ds}\frac{1}{\sqrt{(z - z_0)}}\right\}}{\left(\frac{dF}{dy}\right)},$$

which by means of (1) will become a differential equation of the second order between two variables; the integral will therefore contain two arbitrary constants, which will enable us to make the curve pass through the two given points.

158. PROP. *A particle acted on by any forces, and resting on a smooth horizontal plane, is attached by an inextensible string to a point which moves in a given manner in that plane; to determine the motion of the particle.*

Let x, y, $\bar{x}$, $\bar{y}$ be the co-ordinates at time t of the particle and point, a the length of the string, R the force of constraint, m the mass of the particle.

For the motion of the particle we have

$$\left.\begin{aligned} m\,\frac{d^2x}{dt^2} &= mX - R\,\frac{x-\bar{x}}{a} \\ m\,\frac{d^2y}{dt^2} &= mY - R\,\frac{y-\bar{y}}{a} \end{aligned}\right\} \ldots\ldots(1).$$

with the condition $(x-\bar{x})^2 + (y-\bar{y})^2 = a^2$.

Now $\bar{x}$, y are known functions of t. Take from both sides of the equations in (1) the quantities $m\frac{d^2\bar{x}}{dt^2}$, $m\frac{d^2\bar{y}}{dt^2}$ respectively, and we have the equations of *relative* motion

$$\left.\begin{aligned} m\,\frac{d^2(x-\bar{x})}{dt^2} &= mX - R\,\frac{x-\bar{x}}{a} - m\,\frac{d^2\bar{x}}{dt^2} \\ m\,\frac{d^2(y-\bar{y})}{dt^2} &= mY - R\,\frac{y-\bar{y}}{a} - m\,\frac{d^2\bar{y}}{dt^2} \end{aligned}\right\} \ldots\ldots(2).$$

These are the same equations we should have had if the point had been fixed, and in addition to the forces X, Y and R acting on the particle, we had applied, reversed in direction, the accelerations of the point's motion. It is evident that the same theorem will hold in three dimensions. The accelerations $\frac{d^2\bar{x}}{dt^2}$, $\frac{d^2\bar{y}}{dt^2}$ are known, and therefore the equations of relative motion are completely determined.

159. Let there be no impressed forces, and suppose first that the point moves uniformly in a straight line. Here $\frac{d\bar{x}}{dt}$, $\frac{d\bar{y}}{dt}$ are constant, and therefore no terms are introduced in the equations of motion. We have thus the case of Art. 14.

Or again, suppose the point's motion to be rectilinear, but uniformly accelerated. The relative motion will evidently be that of a simple pendulum from side to side of the point's line of motion.

160. Suppose the point to move with uniform angular velocity ω, in a circle whose radius is r and center origin.

Here supposing the point to start from the axis of x,

$$\bar{x} = r\cos\omega t, \quad \bar{y} = r\sin\omega t.$$

Hence the equations of motion are, since

$$\frac{d^2\bar{x}}{dt^2} = -\omega^2\bar{x}, \quad \frac{d^2\bar{y}}{dt^2} = -\omega^2\bar{y},$$

$$\left.\begin{aligned} \frac{d^2(x-\bar{x})}{dt^2} &= -\frac{R}{m}\cdot\frac{x-\bar{x}}{a} + \omega^2\bar{x}, \\ \frac{d^2(y-\bar{y})}{dt^2} &= -\frac{R}{m}\cdot\frac{y-\bar{y}}{a} + \omega^2\bar{y}, \\ (x-\bar{x})^2 &+ (y-\bar{y})^2 = a^2, \end{aligned}\right\}$$

$$(x-\bar{x})\frac{d^2(y-\bar{y})}{dt^2} - (y-\bar{y})\frac{d^2(x-\bar{x})}{dt^2}$$

$$= \omega^2\{(x-\bar{x})\bar{y} - (y-\bar{y})\bar{x}\};$$

or in polar co-ordinates

$$\frac{d}{dt}\left(a^2\frac{d\theta}{dt}\right) = -\omega^2 ar\sin(\theta-\omega t),$$

$$\text{or} \quad \frac{d^2(\theta-\omega t)}{dt^2} = -\omega^2\frac{r}{a}\sin(\theta-\omega t);$$

or, putting $\phi = \theta - \omega t$, which is the inclination of the string to the radius passing through the point,

$$\frac{d^2\phi}{dt^2} = -\omega^2\frac{r}{a}\sin\phi,$$

which is the ordinary equation of motion of a pendulum.

The particle therefore oscillates about the uniformly revolving radius of the circle described by the point, just as a simple pendulum about the vertical.

161. *Motion of a particle acted on by given forces, and constrained to move in a smooth tube, in the form of a given plane curve, of indefinitely small sectional area, which revolves in a given manner about an axis in its plane.*

Let the axis of revolution be that of z, and let the position of the particle at time t be given by its distance r from that axis, the plane of the tube at that instant making an angle θ with a fixed plane. By the conditions of the problem θ is a given function of t.

The sole effect of the tube will be to produce a force of constraint, which as it lies in the normal plane to the tube may be resolved into two parts, one perpendicular to the plane of the tube, the other in that plane and in the normal to the tube.

Let the impressed forces be resolved into three, P along r, T perpendicular to the plane of the tube and S parallel to the axis of z.

Let R, R' be the two resolved parts of the force of constraint.

The equations of motion will then be (by Arts. 9, 31)

$$m\left\{\frac{d^2r}{dt^2} - r\left(\frac{d\theta}{dt}\right)^2\right\} = mP + R\frac{dz}{ds} \dots\dots\dots\dots (1),$$

$$m\left(r\frac{d^2\theta}{dt^2} + 2\frac{dr}{dt}\cdot\frac{d\theta}{dt}\right) = mT - R' \dots\dots\dots\dots (2),$$

$$m\frac{d^2z}{dt^2} = mS - R\frac{dr}{ds} \dots\dots\dots\dots\dots\dots\dots\dots (3),$$

where s is the arc of the revolving curve.

In addition to these we have the two equations

$$\theta = f(t) \dots\dots\dots\dots\dots\dots\dots\dots\dots (4),$$

which gives the position of the tube at any time, and

$$r = \phi(z) \dots\dots\dots\dots\dots\dots\dots\dots\dots (5),$$

the equation to the tube.

By means of (4) and (5) we may eliminate θ, r, and s from (1), (2), (3). Then eliminating R between (1) and (3), we obtain a differential equation between z and t, whose integral together with (4) completely determines the *position* of the particle at any instant.

R and R' may then be found from (1) or (3), and (2).

In general the angular velocity of the tube is given constant, or $\frac{d\theta}{dt} = \omega$, whence (4) becomes $\theta = \omega t$ if the plane from which θ is measured be that of the tube at the time $t = 0$.

The simplest case we can take is the following.

162. *A particle moves in a straight tube which revolves uniformly round a vertical axis to which it is perpendicular, to determine the motion.*

Here $z =$ constant, $\frac{d\theta}{dt} =$ constant $= \omega$, $P = 0$, and we have from (1)

$$\frac{d^2r}{dt^2} - r\omega^2 = 0;$$

$$\text{whence } r = A\epsilon^{\omega t} + B\epsilon^{-\omega t}.$$

Suppose the motion to commence at time $t = 0$ by the cutting of a string, length r_0, attaching the particle to the axis. The velocity of the particle at that instant along the tube would be zero. Hence at $t = 0$

$$r = r_0 = A + B,$$

$$\frac{dr}{dt} = 0 = A - B;$$

$$\therefore A = B = \frac{r_0}{2};$$

$$\text{and } r = \frac{r_0}{2}(\epsilon^{\omega t} + \epsilon^{-\omega t}).$$

In the figure suppose OM the initial position of the tube, A that of the particle; OL, Q the tube and particle at time t. Then $OA = r_0$, arc $AP = r_0\omega t$, $OQ = r$, and we have

$$OQ = \frac{OA}{2}\left(\epsilon^{\frac{\text{arc } AP}{OA}} + \epsilon^{-\frac{\text{arc } AP}{OA}}\right).$$

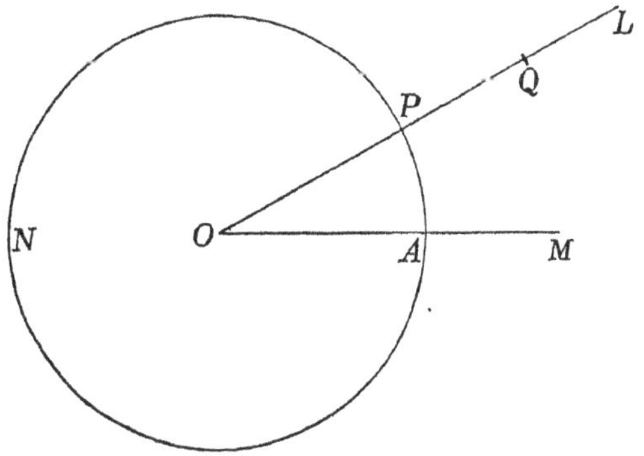

Whence we see that OQ and the arc AP are corresponding values of the ordinate and abscissa of a catenary whose parameter is OA.

Here by (3) we have evidently $R = mg$.

Also by (2), $R' = -2m\left(\frac{r_0\omega}{2}\right)(\epsilon^{\omega t} - \epsilon^{-\omega t})\,\omega$

$$= -m\omega^2 r_0 (\epsilon^{\omega t} - \epsilon^{-\omega t}).$$

From this equation combined with the value of r, we easily deduce

$$R' = 2m\omega^2 \sqrt{(r^2 - r_0^2)},$$

and it is therefore proportional at any instant to the tangent drawn from Q to the circle APN.

163. *Suppose the tube to revolve uniformly in a vertical plane about a horizontal axis.*

We have from equation (1)

$$m\left(\frac{d^2r}{dt^2} - r\omega^2\right) = -mg\cos\omega t,$$

if we conceive the tube to be vertical when $t=0$. Suppose the integral of this equation to be

$$r = A\epsilon^{\omega t} + B\epsilon^{-\omega t} + C\cos\omega t.$$

Differentiating and substituting we find $C = \frac{g}{2\omega^2}$.

$$\text{Hence, } r = A\epsilon^{\omega t} + B\epsilon^{-\omega t} + \frac{g}{2\omega^2}\cos\omega t,$$

$$\text{and if } r = r_0,\ \frac{dr}{dt} = 0, \text{ when } t = 0,$$

$$r_0 = A + B + \frac{g}{2\omega^2},$$

$$0 = A - B;$$

$$\text{or, } r = \left(\frac{r_0}{2} - \frac{g}{4\omega^2}\right)(\epsilon^{\omega t} + \epsilon^{-\omega t}) + \frac{g}{2\omega^2}\cos\omega t,$$

which completely determines the motion. R and R' may be found as before.

164. *Let the tube be in the form of a circle turning uniformly about a vertical diameter.*

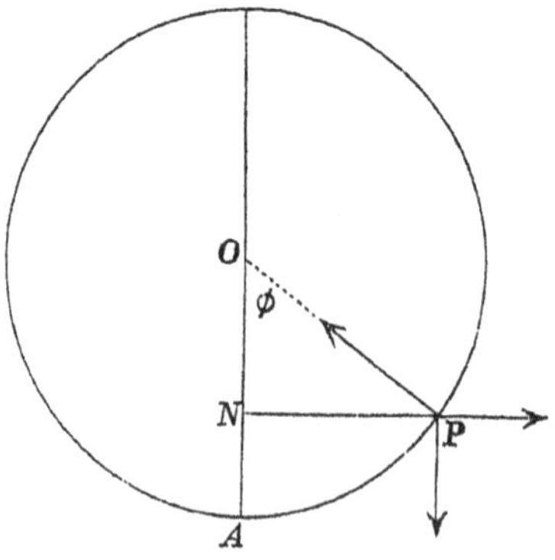

Let AO be the axis, P the position of the particle at any time. Let $POA = \phi$ denote the particle's position. Then the forces acting on it along the element at P are the resolved parts of gravity and of the centrifugal force (Chap. IX.) due to the distance NP from the axis. Hence as $AP = AO \cdot \phi = a\phi$, we have

$$a\frac{d^2\phi}{dt^2} = a\omega^2 \sin\phi \cos\phi - g \sin\phi \dots\dots\dots (1),$$

$$\left(\frac{d\phi}{dt}\right)^2 = \omega^2 \sin^2\phi + \frac{2g}{a}\cos\phi + C.$$

Suppose the particle projected from the lowest point with angular velocity ω_1,

$$\omega_1^2 = \frac{2g}{a} + C,$$

$$\text{and } \left(\frac{d\phi}{dt}\right)^2 = \omega^2\left\{1 - \cos^2\phi - \frac{2g}{a\omega^2}(1 - \cos\phi)\right\} + \omega_1^2.$$

This will be zero if ϕ have a value determined from the equation,

$$\cos^2\phi - \frac{2g}{a\omega^2}\cos\phi = 1 - \frac{2g}{a\omega^2} + \frac{\omega_1^2}{\omega^2};$$

$$\text{or } \cos\phi = \frac{g}{a\omega^2} \pm \sqrt{\left\{\left(1 - \frac{g}{a\omega^2}\right)^2 + \frac{\omega_1^2}{\omega^2}\right\}}.$$

Now as long as $\frac{\omega_1^2}{\omega^2} > \frac{4g}{a\omega^2}$, or $\omega_1^2 > \frac{4g}{a}$, both values are numerically greater than 1, and therefore the motion is one of continuous revolution. If $\omega_1^2 = \frac{4g}{a}$, we have $\frac{d\phi}{dt} = 0$ for $\cos\phi = -1$, and therefore the particle just comes to rest at the highest point. We may notice that $a^2\omega_1^2$, or the velocity of projection from the lowest point is then equal to $2g \cdot 2a$, or is that due to the diameter.

Hence, if a particle be projected from the lowest point of a vertical circle with velocity due to the diameter, it will tend to reach the highest point and there remain at rest whether the circle be fixed or revolving with any angular velocity about the vertical diameter. In this case the position of the particle at any instant can be determined.

If $\omega_1^2 < \frac{4g}{a}$, there is one possible value of $\cos\phi$, and therefore the particle will oscillate about the lowest point.

Suppose again, the projection to be made from the extremity of the horizontal diameter. In this case our corrected equation becomes

$$\left(\frac{d\phi}{dt}\right)^2 = -\omega^2 \cos^2\phi + \frac{2g}{a}\cos\phi + \omega_1^2;$$

and for positions of rest

$$\cos^2\phi - \frac{2g}{a\omega^2}\cos\phi = \frac{\omega_1^2}{\omega^2};$$

$$\text{or } \cos\phi = \frac{g}{a\omega^2} \pm \sqrt{\left(\frac{g^2}{a^2\omega^4} + \frac{\omega_1^2}{\omega_2^2}\right)}.$$

Both values will be numerically less than 1, if

$$\sqrt{\left(\frac{g^2}{a^2\omega^4} + \frac{\omega_1^2}{\omega_2^2}\right)} < 1 - \frac{g}{a\omega^2};$$

$$\text{or } \frac{\omega_1^2}{\omega^2} < 1 - \frac{2g}{a\omega^2},$$

and in this case the oscillations of the particle will be performed between points corresponding to these values of ϕ, and on the same side of the vertical diameter.

165. The position of equilibrium of the particle will be found by putting $\frac{d^2\phi}{dt^2} = 0$. Hence,

$$\cos\phi' = \frac{g}{a\omega^2} \quad \ldots\ldots\ldots\ldots\ldots\ldots\ldots (2).$$

To find the time of a small oscillation about this position, let ψ be the angle of displacement, then by (1),

since $\phi = \phi' + \psi$, ψ being very small,

$$\frac{d^2\psi}{dt^2} = \omega^2 \sin(\phi' + \psi) \left\{\cos(\phi' + \psi) - \frac{g}{a\omega^2}\right\}$$

$$= -\omega^2 \sin^2\phi' . \psi \text{ nearly, by (2),}$$

$$= -\left(\omega^2 - \frac{g^2}{a^2\omega^2}\right)\cdot\psi,$$

and therefore by Art. (53), the time required is

$$\frac{2\pi a\omega}{\sqrt{(a^2\omega^4 - g^2)}} \quad\text{.................... (3).}$$

That there may be a position of equilibrium other than the lowest point, we must have by (2)

$$\omega > \sqrt{\left(\frac{g}{a}\right)},$$

and thus (3) shews that a small oscillation is always possible.

166. Let us find the form of the tube in order that the particle projected with given velocity, may preserve its velocity unchanged, gravity acting parallel to the axis.

Resolving tangentially, and taking co-ordinates x, y in the plane of the curve, y being the axis of revolution,

$$m\frac{d^2s}{dt^2} = mx\omega^2\frac{dx}{ds} - mg\frac{dy}{ds}.$$

$$\text{Hence, } \left.\frac{ds}{dt}\right]^2 = x^2\omega^2 - 2gy + C.$$

$$\text{But } \frac{ds}{dt} = \text{constant.}$$

Hence, $x^2 = \frac{2g}{\omega^2}(y + k)$, the equation to a parabola whose axis is vertical and vertex downwards. This result might easily have been foreseen, as the velocity can only be constant if the accelerating effect of the impressed forces along the curve be zero at every point; and we know that the curve which possesses that property is the parabola.

167. PROP. *A particle moves on a rough curve under the action of given forces; to determine the motion.*

We see (TODHUNTER, *Statics*, Art. 172.) that if μ' be the coefficient of dynamical friction, and $R = \sqrt{(R_1^2 + R_2^2)}$ be the force of constraint as in Art. (144), the effect of friction will be a force $\mu' \sqrt{(R_1^2 + R_2^2)}$ acting in the tangent to the curve, and in the opposite direction to the particle's motion.

Equation (1) of Art. (144), will therefore become

$$\frac{d^2s}{dt^2} = S - \frac{\mu'}{m} \sqrt{(R_1^2 + R_2^2)},$$

the other two equations remaining the same.

If from these three we eliminate R_1 and R_2, we may by means of the equations to the curve eliminate x, y and z, and the final result involving only s and t, suffices to determine the motion completely.

168. Ex. A particle moves in a rough tube in the form of a plane curve, under the action of no forces; to determine the motion.

Here $\frac{d^2s}{dt^2} = -\frac{\mu'}{m} R = -\frac{\mu' v^2}{\rho}$.

Now $v = \frac{ds}{dt}$; $\therefore v \frac{dv}{ds} = \frac{d^2s}{dt^2}$, hence

$$v \frac{dv}{ds} = -\mu' \frac{v^2}{\rho}:$$

$$\text{or } v = a\epsilon^{-\mu' \int \frac{ds}{\rho}}.$$

But if ψ be the angle which the tangent at any point makes with a fixed line

$$\frac{ds}{\rho} = d\psi.$$

Hence, $v = a\epsilon^{-\mu'\psi}$, where a is the velocity when $\psi = 0$.

It may be instructive to compare this result with that for the tension of a string stretched over a rough curve, (TODHUNTER, *Statics*, Art. 196).

If the curve be one of double curvature $\frac{ds}{\rho}$ is the angle between two successive tangents. If the surface containing these be developed, and ϕ represent the angle between the tangents corresponding to the initial and final positions of the particle

$$v = a\epsilon^{-\mu'\phi}.$$

169. PROP. *A particle under the action of given forces moves on a given rough surface; to determine the motion.*

In this case if R be the force of constraint due to the surface, the effect of friction is $\mu' R$ acting in the tangent to the path of the particle, and the equations of Art. (148) become

$$\left.\begin{aligned} \frac{d^2x}{dt^2} &= X + \frac{R}{m}\lambda - \frac{\mu' R}{m}\frac{dx}{ds} \\ \frac{d^2y}{dt^2} &= Y + \frac{R}{m}\mu - \frac{\mu' R}{m}\frac{dy}{ds} \\ \frac{d^2z}{dt^2} &= Z + \frac{R}{m}\nu - \frac{\mu' R}{m}\frac{dz}{ds} \end{aligned}\right\},$$

from which R must be eliminated. The two resulting equations contain x, y, z and t, and if the latter be eliminated, we have one equation in x, y, z which with the equation to the surface, will completely determine the path.

EXAMPLES.

(1) If a particle attached by a string to a point, *just* make complete revolutions in a vertical plane, the tension of the string in the two positions when it is vertical is zero, and six times the weight of the particle, respectively.

(2) A particle projected with given velocity, moves under the action of gravity on a curve in a vertical plane; find the nature of the curve that the pressure on it may be the same throughout the motion.

(3) A particle is projected with given velocity from the vertex of a cycloid whose axis is vertical and vertex uppermost, find where it will leave the curve, and the latus rectum of its future parabolic path.

(4) The axis major of an ellipse is vertical, shew that in order that a particle projected vertically upwards from the extremity of the axis minor along the concave side of the arc may pass through the center after leaving the curve, the velocity of projection must be

$$\left\{\frac{(8a^2+b^2)\,g}{3a\sqrt{(3)}}\right\}^{\frac{1}{2}},$$

a and b being the semiaxes of the ellipse.

(5) The Earth being supposed to be at rest, and to consist of concentric spherical strata with densities varying gradually from the center to the surface, investigate the law of density according to which a particle let fall from the mouth of a diametral pit would perform oscillations exactly similar to those of a simple pendulum oscillating through 45° on each side of the vertical.

(6) Shew that if a particle falling from rest at a point in an inverted cycloid have its velocity suddenly annihilated when it has passed over half its vertical height above the lowest point, and be allowed to proceed always losing its velocity when half way down from the last position of no

velocity, it will be at $\frac{1}{2^{2n}}$th of its original height above the vertex after n times the time it would have taken to fall to the vertex undisturbed.

(7) Shew that a simple pendulum under the action of a central force varying as the distance only, will move as it does under the action of gravity.

(8) The times of oscillation of a pendulum are observed at the Earth's surface, and also at a height h above the surface; from these data find the radius of the Earth supposed spherical.

(9) A pendulum oscillates in a small circular arc, and is acted on in addition to gravity by a small horizontal force as the attraction of a mountain. Shew how to find the latter by observing the number of oscillations gained in a given time. Also find the direction in which the attraction must act so as not to alter the time of oscillation.

(10) (See 2). Determine the nature of the curve about which the string of a simple pendulum must wrap itself in order that its tension may be constant, and deduce the equation between the length of the arc, and the vertical ordinate

$$y = l - \frac{T}{g}(l - s) + C(l - s)^{\frac{2}{3}} - \frac{V^2}{2g},$$

where l is the length of the string, T the constant tension, and V the velocity of the bob when the string is vertical.

(11) A string wrapped round a regular polygon has a particle at the free end which just reaches an angle. There is in the center a repulsive force $\propto (D)$. If v_r be the velocity when r sides are unwrapped, shew that

$$v_r^2 = \frac{r.(r+1)}{2} v_1^2,$$

the particle starting from rest.

(12) Find the curve cutting a series of ellipses with the same vertical axis and vertex, so that a particle descending

each of them from rest at the point of section may press equally at the vertex; in the following cases,

(α) vertical axes $= a$.

(β) horizontal axes $= b$.

(γ) ellipses similar.

(13) Find the equation to a curve in a vertical plane, such that if a particle descend along it, the parts of the pressure due to the velocity, and to gravity, may have a given ratio.

If this be e, and the axis of y be vertical, then the differential equation is

$$\overline{\frac{dy}{dx}}\Big|^2 = \left(\frac{C}{y}\right)^e - 1,$$

where C is a constant.

(14) Find a curve such that a particle starting from rest will describe any arc in the same time as the chord, the force being central and $\propto$ distance.

Deduce (138) as a particular case.

(15) Also find the curve in (138) when the time down the chord is in a given ratio to that down the arc.

(16) Find the curve in which a particle acted on by gravity will revolve uniformly about a point in the same vertical plane.

(17) A particle acted on by a central repulsive force varying as the distance moves in a tube of the form of an epicycloid, the pole being at the center of force. Shew that the oscillations are isochronous.

(18) A particle is initially at rest at a point of the spiral $r = c\epsilon^{-m\theta}$, distant d from the pole. Shew that if the pole be a center of force whose attraction $= \frac{\mu}{D^2}$, the time of fall to it is

$$\frac{\pi}{2} \cdot \frac{d^{\frac{3}{2}}}{\sqrt{(2\mu)}} \sqrt{\left(1 + \frac{1}{m^2}\right)}.$$

(19) In the preceding problem find the pressure on the curve at any instant.

(20) A particle starts from rest at any point of the convex circumference of a vertical circle, shew that it will leave the circle after descending one-third of its original vertical height above the center.

(21) A particle under the influence of gravity, is projected from one point in a horizontal direction towards another point, find the curve on which it must be constrained to move so as to approach uniformly to the latter point.

(22) Two particles are projected from the same point, in the same direction, and with the same velocity, but at different instants, in a smooth circular tube of small bore whose plane is vertical, to shew that the line joining them constantly touches another circle.

Let the tube be called the circle A, and the horizontal line, to a fall from which the velocity is due, L. Let m, m' be corresponding positions of the particles. Suppose that mm' passes into its next position by turning about O, these two lines will intercept two indefinitely small arcs at m and m', which (by a property of the circle) are in the ratio $mO : m'O$.

Let another circle B be described touching mm' in O, and such that L is the radical axis of A and B. Let a be the distance between their centers, mp, $m'p'$ perpendiculars on L. Let mp cut A again in q and B in r, s.

Then by Geometry,

$$mO^2 = rm \,.\, ms = pm\,(rm - qs) = 2a \,.\, pm = \frac{a}{g}\,(\text{velocity of } m)^2.$$

Similarly,

$$m'O^2 = 2a \,.\, p'm' = \frac{a}{g}\,(\text{velocity of } m')^2.$$

Hence the velocities of m and m' are as $mO : m'O$, and therefore by what we have shewn above about elementary arcs at m and m', the proximate position of mm' is also a tangent to B, which proves the proposition.

It is easily seen from this, that if *one* polygon of a given number of sides can be inscribed in one circle and circumscribed about another, an indefinite number can be drawn. For this we have only to suppose a number of particles moving in A with velocities due to a fall from L, if they form at any time the angular points of a polygon whose sides touch B, they will continue to do so throughout the motion. This however does not belong to our subject.

(23) A particle under the action of gravity is projected with given velocity from a point, find the curve on which it must be constrained to move so as to recede uniformly from the point of projection.

(24) A railway train travels due north with given velocity. Compare the horizontal pressure on the rails due to the Earth's rotation, with the weight of the train.

(25) A particle attached by a string to a point moves in a horizontal plane. A small ring passing round the string moves uniformly in a straight line from the point. Shew how to find the equation to the actual path, and shew that that relative to the ring has the equation

$$\rho(1+e\theta)=C.$$

(26) A particle descends from rest under the action of gravity. Find the curve on which it must move in order that the ratio of the times of descending two vertical spaces whose ratio is given, may also be equal to a given quantity.

Verify the general result in the particular cases:

(α) Double the height in double the time.

(β) Four times the height in eight times the time.

(27) s is the arc, and y the vertical ordinate of a curve passing through the origin. If time through s : time through chord of s :: ks : chord, shew that

$$s=Ay^{\frac{k}{2k-1}}.$$

(28) Given an arc of a curve, find the position in which it must be fixed, that a particle starting from rest may describe it under the action of gravity in the least time. Apply the result to an arc of the cardioide measured from the cusp.

(29) A series of similar and similarly situated curves start from a point A in a vertical plane; to find the synchronous curve, or that which cuts off from each of the series a portion which would be described in a given time τ by a particle starting from rest at A.

Taking A as origin, and the axis of y vertically downwards, we have

$$\tau = \int_0^y \sqrt{\left\{\frac{1+\left(\frac{dx}{dy}\right)^2}{2gy}\right\}} dy \ldots\ldots\ldots (1).$$

Now the common equation of the curves contains *one* arbitrary parameter α, which will therefore appear in (1) when x is eliminated. Supposing then that (1) can be integrated, if between the integral and the equation to the curves we eliminate α, the result will be the equation to the required series of synchronous curves, in which τ will appear as a variable parameter.

The equation to the given series of curves will of course be in the form $\frac{x}{\alpha} = f\left(\frac{y}{\alpha}\right)$, so that if $\frac{y}{\alpha} = \omega$, $\frac{dx}{dy}$ in terms of ω will not involve α. It is thence easy to deduce the following values of x and y for the required curve, in terms of ω,

$$x = \frac{\tau^2 f(\omega)}{\left[\int_0^\omega \sqrt{\left\{\frac{1+\left(\frac{dx}{dy}\right)^2}{2g\omega}\right\}} d\omega\right]^2},$$

$$y = \frac{\tau^2 \omega}{\left[\int_0^\omega \sqrt{\left\{\frac{1+\left(\frac{dx}{dy}\right)^2}{2g\omega}\right\}} d\omega\right]^2};$$

from which, if (1) is not integrable, the required curve can be constructed by quadratures.

(30) If t be the time in a cycloid from the point whose abscissa is the radius of the generating circle, to any other point; τ the time down the chord of the generating circle corresponding to the same two points, shew that

$$2 \tan^{-1} \tau \sqrt{\left(\frac{g}{2}\right)} = \tan^{-1} \left\{\sqrt{(2)} \tan t \sqrt{\left(\frac{g}{2}\right)}\right\}.$$

(31) A particle moves in a circular groove radius a under the action of a center of force $\propto D^{-2}$ situated at a distance b from the center of the circle. It is projected from the nearest point with velocity V, shew that for a complete revolution

$$V^2 \nless \frac{4\mu b}{a^2 - b^2}.$$

(32) A particle acted on by gravity descends from rest at a given point, find the nature of the curve on which it must move that the pressure may be at any instant as the square of the vertical height fallen through.

(33) Find the tautochrone when the force is as the cube root of the distance from the axis of x, and parallel to that of y.

$$x^{\frac{2}{3}} + y^{\frac{2}{3}} = a^{\frac{2}{3}}.$$

(34) To find all the tautochrones when the force is central, and varies as the distance.

If S be the force resolved along the curve, we must have

$$\frac{dS}{ds} = -\frac{\pi^2}{4\tau^2} \text{ (Art. 146).}$$

Now if ϕ be the angle between the radius vector and tangent,

$$S = -\mu r \cos\phi = -\mu r \frac{dr}{ds};$$

$$\therefore \frac{d\left(r\dfrac{dr}{ds}\right)}{ds} = \frac{\pi^2}{4\mu\tau^2};$$

$$\therefore r\frac{dr}{ds}\frac{d}{ds}\left(r\frac{dr}{ds}\right) = \frac{\pi^2}{4\mu\tau^2} r\frac{dr}{ds},$$

$$\text{or } \left(r\frac{dr}{ds}\right)^2 = \frac{\pi^2}{4\mu\tau^2} r^2 + C;$$

$$\text{that is, } r^2 - p^2 = \frac{\pi^2}{4\mu\tau^2} r^2 + C.$$

And if a be that perpendicular from the center on the tangent which meets the latter at its point of contact, the corresponding value of r is a also, and therefore

$$C = -\frac{\pi^2}{4\mu\tau^2} a^2;$$

$$\therefore p^2 = \left(1 - \frac{\pi^2}{4\mu\tau^2}\right) r^2 + \frac{\pi^2}{4\mu\tau^2} a^2 \ldots\ldots\ldots\ldots\ldots (1),$$

which is the differential equation to the required curves.

The curves will differ in species according to the value of $\frac{\pi^2}{4\mu\tau^2}$.

I. Let $\frac{\pi^2}{4\mu\tau^2} > 1$, (1) takes the form

$$p^2 = e^2 (c^2 - r^2),$$

which is at once recognized as the differential equation to the epicycloid traced by a point in the circumference of a small circle, which rolls on the inner surface of a large one whose center is the pole.

If we suppose the radius of the larger circle to be indefinitely increased, its circumference tends to become a straight line and the epicycloid to become a series of cycloids. The force in this case tends to be constant, and perpendicular to the bases of these cycloids, whence in the limit we have the result of Art. (131).

II. Let $\frac{\pi^2}{4\mu\tau^2} = 1$.

Here $p = a$ and the curve is a straight line. This is the case of Art. (48).

III. $\frac{\pi^2}{4\mu\tau^2} < 1$. In this, as in the other cases, we may find the equation to the curve by integrating (1) after substituting for p its value in terms of r and θ, but the equation we thus obtain is very complicated. This curve is found to be a spiral with two symmetrical branches extending to an infinite distance, and approximating to spirals which make angles $= \cos^{-1} e$ with the radius vector.

IV. Let the origin be on the curve, and be the point to which the time of descent is measured. Then $a = 0$.

If $\frac{\pi^2}{4\mu\tau^2} > 1$, the equation is impossible.

If it $= 1$, we have a straight line through the origin.

If it < 1, we have the Logarithmic Spiral.

V. If the force be repulsive, we require only to change the sign of μ. In this case we have an epicycloid traced by a point of a circle which rolls on the outside of another whose center is origin. From this again we may deduce the tautochronism of the common cycloid for gravity.

(35) A particle moving on the interior surface of a vertical circular cylinder is projected with a given velocity, and goes round n times before it falls to the level of the point of projection. Determine the direction of projection.

(36) Shew that a particle moving under the action of gravity on a smooth helix whose axis is vertical, makes the first revolution from rest in the time

$$\sqrt{\left(\frac{8\pi a}{g \sin 2\alpha}\right)}.$$

(37) A groove is cut on a right cone of height h, at an angle β with the generating line. Shew that the time of reaching the base, from a vertical height h_1 below the vertex, by a particle sliding in the groove is

$$\frac{\sqrt{h} - \sqrt{h_1}}{\sqrt{(8g)} \cdot \cos\alpha \cos\beta},$$

where α is the semivertical angle.

(38) Find the curve on the surface of a vertical cylinder down which if a particle slide, the force of constraint will be constant.

(39) A particle moves on a smooth ellipsoid, under the action of a force $\propto (D)$ in the center. Given the velocity and direction with which it passes the extremity of an axis, find the pressure.

(40) A smooth tube of indefinitely small bore revolves in a horizontal plane. A particle attached to the axis by an elastic string moves in the tube. Determine the conditions that the motion be oscillatory.

(41) A circular tube of indefinitely small bore revolves with uniform angular velocity ω about a vertical diameter, and a particle in it is projected from the lowest point with velocity due to the diameter. Determine the motion, and shew that it is at its greatest distance from the axis after a time

$$\left(\frac{g}{a} + \omega^2\right)^{-\frac{1}{2}} \log_e \sqrt{\left\{\frac{\sqrt{\left(\frac{2g}{a} + \omega^2\right)} + \sqrt{\left(\frac{g}{a} + \omega^2\right)}}{\sqrt{\left(\frac{2g}{a} + \omega^2\right)} - \sqrt{\left(\frac{g}{a} + \omega^2\right)}}\right\}};$$

where a is the radius of the tube.

(42) The axis of a rough helix of radius a is vertical, and the curve makes an angle β with the horizon; a ring slides on it with initial velocity

$$\frac{\sqrt{(ga)}\,(\sin^2\beta - \mu^2 \cos^2\beta)^{\frac{1}{2}}}{\sqrt{(\mu)} \cdot \cos\beta},$$

determine the motion.

(43) A heavy particle attached to a point by a string whose unstretched length is a, lies on a rough horizontal plane and is projected perpendicular to the string with velocity v. If it comes to rest at a distance a from the point, after describing a distance s, $v^2 = 2\mu g s$.

(44) A particle descends a rough circular tube from the extremity of the horizontal diameter. If it stops at the lowest point, shew that

$$3\mu\epsilon^{-\mu\pi} + 2\mu^2 = 1.$$

(45) Shew that the result of Art. (150) is true if the surface be rough. If a particle be projected with velocity V along the inner surface of a rough sphere, determine the motion, and shew that it will return to the point of projection in the time

$$\frac{r}{\mu' V}(\epsilon^{2\mu'\pi} - 1),$$

where r is the radius of the sphere.

(46) If the only impressed force be a central one $\propto \frac{\mu}{D^n}$, and the velocity be that from infinity, shew that the equation to the brachistochrone is

$$r^{\frac{n+1}{2}} \cos \frac{n+1}{2}(\theta + \beta) = a^{\frac{n+1}{2}}.$$

(47) A material particle P, attached by a slender cord of given length a, to a point S in a fixed axis SA, is attracted by a constant force g in a direction always parallel to a line SB, which is inclined at a given angle to the axis SA, and revolves about it with a given angular velocity ω: shew that if $V =$ the velocity of P, $\omega' =$ the angular velocity of the plane PSA about SA, $\phi = \angle PSB$, $\theta = \angle PSA$,

$$V^2 = 2ga \cos\phi + 2a^2\omega\omega' \sin^2\theta + \text{const.}$$

Shew also that the dynamical conditions of this Problem are the same as those of a ball-pendulum acted upon by gravity, when the Earth's rotation is taken into account.

CHAPTER VIII.

MOTION IN A RESISTING MEDIUM.

170. WHEN a body moves in a fluid, whether incompressible or aeriform, it must in displacing the particles of the medium lose part of its own velocity. The resistance of a fluid therefore to a body moving in it is of the nature of a retarding force; but in consequence of the great difficulty of making accurate experiments on the subject, the laws of the resistance of fluids have not yet been very satisfactorily ascertained.

For a velocity neither very great nor very small, the general approximate law seems to be, that the resistance on a plane surface moving with its plane at right angles to the line of motion is proportional to the extent of the surface, the density of the resisting medium, and the square of the velocity taken conjointly. We are here however only treating of the motion of a particle in which the extent of surface has no place in our consideration, and will assume that the resistance depends entirely on the density of the medium and the velocity of the particle, illustrating by supposing different laws the method of procedure in all such cases.

171. PROP. *A particle acted on by no forces is projected in a resisting medium of uniform density, of which the resistance varies as the nth power of the velocity; to determine the motion.*

The motion will evidently be rectilinear. Let x be the distance of the particle from a fixed point in the line of motion at the time t, v its velocity at that time; then the force due to the resistance may be represented by Kv^n, K being a constant, and the equation of motion becomes

$$m\frac{d^2x}{dt^2} = -Kv^n,$$

$$\text{or } \frac{dv}{dt} = -kv^n \text{ .... } (1);$$

if we put $K = mk$, k being of $1-n$ dimensions.

$$\therefore\ k\frac{dt}{dv} = -v^{-n};$$

$$\therefore\ kt = C + \frac{v^{1-n}}{n-1}.$$

Suppose the velocity of projection V, and the particle projected from the origin, then when

$$t = 0,\ v = V;\ \therefore\ C = -\frac{V^{1-n}}{n-1};$$

$$\therefore\ (n-1)\,kt = v^{1-n} - V^{1-n} \ldots\ldots\ldots\ldots\ldots (2),$$

the relation between v and t. It shews that v can never be zero if n is > 1, but if n is < 1 the velocity will become zero when $t = \frac{V^{1-n}}{(1-n)\,k}$. After this the particle will evidently remain at rest.

To find the distance of the particle from the origin at any time, we have from (2)

$$v \text{ or } \frac{dx}{dt} = \{V^{1-n} - (1-n)\,kt\}^{\frac{1}{1-n}};$$

$$\therefore\ x = -\frac{1}{(2-n)\,k}\{V^{1-n} - (1-n)\,kt\}^{\frac{2-n}{1-n}} + \frac{V^{2-n}}{(2-n)\,k};$$

x and t being supposed to vanish together. Hence when n is < 1, the distance to which the particle will go, or its distance at the time $\frac{V^{1-n}}{(1-n)\,k}$

$$= \frac{V^{2-n}}{(2-n)\,k}.$$

172. There is one case in which the above solution fails, namely when $n = 1$, or the resistance varies as the velocity. In this case, by (1)

$$\frac{dv}{dt} = -kv, \quad \therefore k\frac{dt}{dv} = -\frac{1}{v},$$

$$\therefore kt = C - \log v.$$

When $t = 0$, $v = V$; $\therefore C = \log V$;

$$\therefore kt = \log \frac{V}{v} \quad \ldots\ldots\ldots\ldots\ldots\ldots\ldots (3);$$

$$\therefore v \text{ or } \frac{dx}{dt} = V\epsilon^{-kt}.$$

Integrating $$x = \frac{V}{k}(1 - \epsilon^{-kt}) \quad \ldots\ldots\ldots\ldots\ldots\ldots (4),$$

the constant being determined as before that x and t may vanish together.

Equations (3) and (4) determine the velocity and position of the particle at any instant, and shew that the velocity continually diminishes without ever actually becoming zero, but that the space passed over by the particle can never be greater than a certain quantity, for when

$$t = \infty, \quad x = \frac{V}{k}.$$

173. Prop. *A particle acted on by a constant accelerating force in its line of motion, moves in a resisting medium of uniform density, of which the resistance varies as the square of the velocity; to determine the motion.*

Suppose the particle projected from the origin with the velocity V, and let v be its velocity at any time t, x its distance at that time, and f the constant accelerating force.

Assume K to be the velocity with which the particle would have to be animated that the resisting force might be equal to mf, then the retarding force at any time may be represented by $mf\,\frac{v^2}{K^2}$.

I. Let f act to diminish x; then the equation of motion is

$$\frac{d^2x}{dt^2} = -f - f\frac{v^2}{K^2},$$

which since

$$\frac{d^2x}{dt^2} = \frac{dv}{dt} = \frac{dv}{dx}\,\frac{dx}{dt} = v\frac{dv}{dx}$$

may be written either

$$\frac{dv}{dt} = -\frac{f}{K^2}(K^2 + v^2)$$

$$\text{or } v\frac{dv}{dx} = -\frac{f}{K^2}(K^2 + v^2).$$

These again may be changed into

$$\frac{dt}{dv} = -\frac{K^2}{f}\cdot\frac{1}{K^2 + v^2},$$

$$\frac{dx}{d(v^2)} = -\frac{K^2}{2f}\cdot\frac{1}{K^2 + v^2}.$$

Integrating, and determining the constants so that when

$$x = 0, \quad t = 0, \quad v = V,$$

we obtain

$$\frac{ft}{K} = \tan^{-1}\frac{V}{K} - \tan^{-1}\frac{v}{K} = \tan^{-1}\frac{K(V - v)}{K^2 + Vv},$$

$$\frac{2fx}{K^2} = \log\frac{K^2 + V^2}{K^2 + v^2}.$$

Let T be the time at which the velocity becomes zero, and h the corresponding value of x, then

$$T = \frac{K}{f}\tan^{-1}\frac{V}{K}, \text{ and } h = \frac{K^2}{2f}\log\left(1 + \frac{V^2}{K^2}\right).$$

After this the particle begins to return, the force of resistance therefore tends to increase x, and the equation of motion is

$$\frac{d^2x}{dt^2} = -f + f\frac{v^2}{K^2},$$

which as before may be written either

$$-\frac{dv}{dt} = -\frac{f}{K^2}(K^2 - v^2)$$

$$\text{or } v\frac{dv}{dx} = -\frac{f}{K^2}(K^2 - v^2).$$

These may be changed into

$$\frac{dt}{dv} = \frac{K^2}{f} \cdot \frac{1}{K^2 - v^2},$$

$$\frac{dx}{d(v^2)} = -\frac{K^2}{2f} \cdot \frac{1}{K^2 - v^2}.$$

Integrating and determining the constants, so that when

$$v = 0, \quad x = h, \quad t = T,$$

we obtain

$$\frac{2f}{K}(t - T) = \log\frac{K + v}{K - v},$$

$$\frac{2f}{K^2}(h - x) = \log\frac{K^2}{K^2 - v^2}.$$

Let U be the velocity with which the particle will return to the point of projection, then putting $x = 0$ in the latter equation, we obtain

$$\frac{U^2}{K^2} = 1 - \epsilon^{-\frac{2fh}{K^2}};$$

or substituting for h its value,

$$\frac{U^2}{K^2} = \frac{\frac{V^2}{K^2}}{1 + \frac{V^2}{K^2}},$$

whence

$$\frac{1}{U^2} - \frac{1}{V^2} = \frac{1}{K^2}.$$

II. Let f act to increase x. Then

$$\frac{d^2x}{dt^2} = f - f\frac{v^2}{K^2},$$

from which we derive, as before,

$$\frac{dt}{dv} = \frac{K^2}{f} \cdot \frac{1}{K^2 - v^2},$$

$$\frac{dx}{d(v^2)} = \frac{K^2}{2f} \cdot \frac{1}{K^2 - v^2}.$$

Integrating and determining the constants, that when

$$t = 0, \quad x = 0, \quad v = V,$$

we obtain

$$t = \frac{K}{2f} \cdot \log \frac{(K+v)(K-V)}{(K-v)(K+V)},$$

$$x = \frac{K^2}{2f} \cdot \log \frac{K^2 - V^2}{K^2 - v^2}.$$

From the latter equation we obtain

$$v^2 = K^2 - (K^2 - V^2)\,\epsilon^{-\frac{2fx}{K^2}}.$$

This equation shews that when x becomes very large, v approaches to K, which is its limiting value. If the velocity of projection be less than K, v will continually approach to K, and never exceed it, and if the velocity be greater than K, v will constantly diminish towards K, and never become less.

174. The results of the last Proposition will apply to bodies projected in a resisting medium vertically upwards or downwards under the action of gravity; for the accelerating force

of gravity may still be considered constant although not the same as for a particle in vacuo. The actual moving force is in fact the difference of the weights of the body and the fluid displaced, so that if α be the ratio of the specific gravity of the fluid displaced to that of the body, the moving force

$$= W(1-\alpha) = Mg(1-\alpha),$$

where W and M are the weight and mass of the body, and therefore the accelerating force arising from it $= g(1-\alpha)$. By substituting this for f in the results of Art. (173) we may obtain formulæ for the motion of bodies in a vertical direction under the action of gravity.

175. Prop. *To find the equations of motion in a resisting medium of a particle acted on by any forces.*

Let x, y, z be the co-ordinates of the particle relative to an assumed system of rectangular axes, at the time t, and let X, Y, Z be the component accelerating effects, parallel to the axes, of the forces acting on the particle. Then denoting by R the accelerating effect of the resistance, which acts in the tangent to the path described and in a direction opposed to the motion, we have

$$\frac{d^2x}{dt^2} = X - R\frac{dx}{ds},$$

$$\frac{d^2y}{dt^2} = Y - R\frac{dy}{ds},$$

$$\frac{d^2z}{dt^2} = Z - R\frac{dz}{ds}.$$

These are the general equations of motion. In any particular case R will be given as a function of the density and velocity, and particular methods will be necessary for obtaining the path of the particle and its position at any time. The method of procedure will be illustrated in what follows.

176. Prop. *A particle acted on by a constant force parallel to a fixed line is projected from a given point in a given direction with a given velocity, and moves in an uniform medium whose resistance varies as the square of the velocity; to determine the motion.*

Take the given point as origin, the axis of x perpendicular, and that of y parallel, to the given line. Let f be the constant accelerating force, acting we will suppose to diminish y; then the accelerating effect of the resistance may be represented by $k\left(\frac{ds}{dt}\right)^2$, where k is a constant. Hence the equations of motion are

$$\frac{d^2x}{dt^2} = -k\left(\frac{ds}{dt}\right)^2 \frac{dx}{ds} \quad \text{.................. (1)},$$

$$\frac{d^2y}{dt^2} = -f - k\left(\frac{ds}{dt}\right)^2 \frac{dy}{ds} \quad \text{........... (2)}.$$

The former may be written

$$\frac{d^2x}{dt^2} = -k\frac{ds}{dt}\frac{dx}{dt};$$

and therefore dividing by $\frac{dx}{dt}$ and integrating,

$$\log\frac{dx}{dt} = C - ks.$$

Suppose u the component of the initial velocity parallel to x, then when $s = 0$,

$$\frac{dx}{dt} = u; \quad \therefore\ C = \log u, \text{ and}$$

$$\therefore\ \frac{dx}{dt} = u\epsilon^{-ks}.$$

Hence $\frac{dy}{dt} = \frac{dy}{dx}\frac{dx}{dt} = u\epsilon^{-ks}\frac{dy}{dx}$;

$$\therefore\ \frac{d^2y}{dt^2} = u\epsilon^{-ks}\frac{d^2y}{dx^2}\frac{dx}{dt} - ku\epsilon^{-ks}\frac{ds}{dt}\frac{dy}{dx}$$

$$= u^2\epsilon^{-2ks}\frac{d^2y}{dx^2} - ku\epsilon^{-ks}\frac{ds}{dt}\frac{dy}{dx}.$$

But by (2),

$$\frac{d^2y}{dt^2} = -f - k\frac{ds}{dt}\frac{dy}{dt}$$

$$= -f - ku\epsilon^{-ks}\frac{ds}{dt}\frac{dy}{dx}.$$

Comparing these we obtain

$$\frac{d^2y}{dx^2} + \frac{f}{u^2}\epsilon^{2ks} = 0;$$

or putting $p = \dfrac{dy}{dx}$,

$$\frac{dp}{dx} + \frac{f}{u^2}\epsilon^{2ks} = 0 \quad\ldots\ldots\ldots\ldots\ldots\ldots\ldots (3),$$

the differential equation to the path.

It may be made integrable once by multiplying it by

$$\sqrt{(1+p^2)} = \frac{ds}{dx},$$

which gives

$$\frac{dp}{ds}\sqrt{(1+p^2)} + \frac{f}{u^2}\epsilon^{2ks} = 0.$$

Integrating and determining the constant so that when $s = 0$, $p = \tan\alpha$, α being the angle which the direction of projection makes with the axis of x, we obtain

$$p\sqrt{(1+p^2)} + \log\{p + \sqrt{(1+p^2)}\} + \frac{f}{ku^2}\epsilon^{2ks}$$

$$= \tan\alpha\sec\alpha + \log(\tan\alpha + \sec\alpha) + \frac{f}{ku^2} \ldots\ldots\ldots (4).$$

The equation cannot be integrated farther.

If we make $k = 0$ in equations (1) (2), we have the equations which belong to the trajectory in a non-resisting medium, the original velocity and direction of projection being the same as in this problem. Hence if S, s be arcs of the trajectories in a non-resisting and resisting medium measured from the point of projection to any two points at which the tangents are parallel,

$$\frac{dp}{ds}\sqrt{(1+p^2)} = -\frac{f}{u^2}\frac{dS}{ds} = -\frac{f}{u^2}\epsilon^{2ks};$$

$$\therefore\ \frac{dS}{ds} = \epsilon^{2ks};$$

$$\therefore\ 2kS = \epsilon^{2ks} - 1,$$

since we suppose S and s to commence together.

$$\text{Hence } 2ks = \log(1 + 2kS) \quad\ldots\ldots\ldots\ldots\ldots (5).$$

177. From equation (3) it appears that as s becomes more and more nearly equal to $+\infty$, $\frac{dx}{dp}$ becomes always more and more nearly zero, and therefore x becomes more and more nearly equal to a constant. Hence the curve on the positive side of the origin tends continually to coincide with a straight line parallel to the axis of y, at a finite distance, which is therefore an asymptote.

Again, as s becomes more and more nearly equal to $-\infty$, $\frac{dp}{dx}$ becomes more and more equal to zero, or p tends to become constant.

The curve therefore on the negative side of the origin tends to become parallel to a certain straight line. It appears also from equation (5) that when $S = -\frac{1}{2k}$, s becomes $-\infty$, and therefore the curve tends to become parallel to the tangent at a point of the common parabola at a distance $-\frac{1}{2k}$ along the curve from the origin. Or if θ be the angle which this straight line makes with the axis of x, we have by putting $s = -\infty$ in equation (4),

$$\tan\theta\sec\theta - \tan\alpha\sec\alpha + \log\frac{\tan\theta + \sec\theta}{\tan\alpha + \sec\alpha} = \frac{f}{ku^2}.$$

To shew that the curve has an asymptote parallel to this line, we must prove that $\bar{x}$ the distance from the origin, of the intersection of the tangent with the axis, is always finite.

$$\text{Now } \bar{x} = x - y\frac{dx}{dy};$$

$$\therefore \frac{d\bar{x}}{dx} = -y\frac{d}{dx}\cdot\frac{dx}{dy} = +\frac{y}{p^2}\frac{dp}{dx}.$$

Also if the ultimate value of p be called n we shall have ultimately,

$$y = nx, \quad s = x\sqrt{(1+n^2)},$$

and by (3),

$$\frac{dp}{dx} = -\frac{f\epsilon^{2ks}}{u^2};$$

$$\therefore \frac{d\bar{x}}{dx} = -\frac{fx}{nu^2}\epsilon^{2kx\sqrt{(1+n^2)}};$$

$$\therefore \bar{x} = -\int_0^x \frac{fx}{nu^2}\epsilon^{2kx\sqrt{(1+n^2)}}\,dx,$$

which by integration will be found to be finite when x is infinite, and negative.

Hence the curve is not similar on both sides of the vertex. The particle rises more obliquely and descends more vertically than it would do in a vacuum.

178. The projectile will have reached the highest point when $\frac{dy}{dx} = 0$, which gives for the length of the path between the origin and the highest point, by equation (4),

$$\frac{1}{2k}\log\left[1 + \frac{ku^2}{f}\{\tan\alpha\sec\alpha + \log(\tan\alpha + \sec\alpha)\}\right],$$

and for the velocity there

$$u\left[1+\frac{ku^2}{f}\{\tan\alpha\sec\alpha+\log(\tan\alpha+\sec\alpha)\}\right]^{-\frac{1}{2}}.$$

179. The above results will as in Art. (174) be made applicable to the motion of a body projected in the air under the action of gravity, by writing $g(1-\alpha)$ for f. The most important application of the problem is to the practice of gunnery, in which case p is always small, and an approximate equation to the path may be found.

For we have

$$\frac{dp}{dx}+\frac{f}{u^2}\epsilon^{2ks}=0.$$

Multiplying this by $\frac{ds}{dx}=\sqrt{(1+p^2)}=1$, neglecting higher powers of p than the first, we have

$$\frac{dp}{dx}+\frac{f}{u^2}\epsilon^{2ks}\frac{ds}{dx}=0.$$

Integrating and observing that when $s=0$, $p=\tan\alpha$,

$$p+\frac{f}{2ku^2}\epsilon^{2ks}=\tan\alpha+\frac{f}{2ku^2};$$

$$\therefore\ \epsilon^{2ks}=1+\frac{2ku^2}{f}(\tan\alpha-p).$$

Substituting in equation (3),

$$\frac{dp}{dx}+\frac{f}{u^2}+2k(\tan\alpha-p)=0,$$

$$\text{or}\ \frac{dp}{dx}-2kp=-\left(\frac{f}{u^2}+2k\tan\alpha\right).$$

Multiply by ϵ^{-2kx} and integrate, determining the constant as before;

$$\therefore\ p\epsilon^{-2kx}=-\frac{f}{2ku^2}+\left(\frac{f}{2ku^2}+\tan\alpha\right)\epsilon^{-2kx};$$

$$\therefore\ p=\frac{dy}{dx}=-\frac{f}{2ku^2}\epsilon^{2kx}+\left(\frac{f}{2ku^2}+\tan\alpha\right).$$

Integrating and observing that when $x=0,\ y=0$,

$$y=x\left(\tan\alpha+\frac{f}{2ku^2}\right)-\frac{f}{4k^2u^2}(\epsilon^{2kx}-1),$$

the approximate equation to the required path.

180. Prop. *A particle moving in a resisting medium is acted on by a force which is constantly parallel to a fixed line; to find the resistance that a given curve may be described.*

Taking the fixed line as the axis of y, and denoting the force at any point by Y and the resistance by R, the equations of motion will be

$$\frac{d^2x}{dt^2}=-R\frac{dx}{ds}\ \ldots\ldots\ldots\ldots\ldots\ldots\ldots\ldots \ (1),$$

$$\frac{d^2y}{dt^2}=Y-R\frac{dy}{ds}\ \ldots\ldots\ldots\ldots\ldots\ldots\ldots \ (2);$$

$\therefore$ eliminating R,

$$\frac{dx}{dt}\frac{d^2y}{dt^2}-\frac{dy}{dt}\frac{d^2x}{dt^2}=Y\frac{dx}{dt}\ \ldots\ldots\ldots\ (3).$$

$$\text{Now}\ \frac{dy}{dx}=\frac{\frac{dy}{dt}}{\frac{dx}{dt}};$$

$$\therefore\ \frac{d^2y}{dx^2}=\frac{\frac{dx}{dt}\frac{d^2y}{dt^2}-\frac{dy}{dt}\frac{d^2x}{dt^2}}{\left(\frac{dx}{dt}\right)^3},$$

$$=\frac{Y}{\left(\frac{dx}{dt}\right)^2};$$

$$\therefore\ \left(\frac{dx}{dt}\right)^2=\frac{Y}{\frac{d^2y}{dx^2}}\ \ldots\ldots\ldots\ldots\ldots\ldots\ldots\ldots \ (4).$$

Differentiating with respect to t,

$$2\frac{dx}{dt}\frac{d^2x}{dt^2}=\frac{d}{dx}\left(\frac{Y}{\frac{d^2y}{dx^2}}\right)\frac{dx}{dt};$$

$$\therefore \frac{d^2x}{dt^2}=\frac{1}{2}\frac{d}{dx}\left(\frac{Y}{\frac{d^2y}{dx^2}}\right);$$

$\therefore$ by equation (1)

$$R=-\frac{1}{2}\frac{ds}{dx}\cdot\frac{d}{dx}\left(\frac{Y}{\frac{d^2y}{dx^2}}\right)\text{..............} (5).$$

From this equation if we know Y, the resistance that a given curve may be described can be found.

181. If the resistance vary as the product of the density into the nth power of the velocity, equations (4) and (5) may be used to find the law of density that a given curve may be described. For in this case R will be represented by $k\left(\frac{ds}{dt}\right)^n$, where k is proportional to the density at any point, and is to be determined as a function of x and y. We have then from (5)

$$k\left(\frac{ds}{dx}\right)^n\left(\frac{dx}{dt}\right)^n=-\frac{1}{2}\frac{ds}{dx}\cdot\frac{d}{dx}\left(\frac{Y}{\frac{d^2y}{dx^2}}\right);$$

$\therefore$ by (4)

$$k=-\frac{1}{2}\left(\frac{ds}{dx}\right)^{1-n}\left(\frac{Y}{\frac{d^2y}{dx^2}}\right)^{-\frac{n}{2}}\frac{d}{dx}\left(\frac{Y}{\frac{d^2y}{dx^2}}\right),$$

which according as n is not or is equal to 2, may be written

$$k=\frac{1}{n-2}\left(\frac{ds}{dx}\right)^{1-n}\frac{d}{dx}\left(\frac{Y}{\frac{d^2y}{dx^2}}\right)^{\frac{2-n}{2}}\text{...........} (6);$$

$$\text{or } k = -\frac{1}{2}\frac{d}{ds}\log\left(\frac{Y}{\frac{d^2y}{dx^2}}\right) \quad\text{.................. (7).}$$

182. Equations (5) and (6) may also clearly be used to determine the laws of force by which a given curve may be described, the density being supposed known.

From equation (3), we obtain

$$\frac{\left(\frac{ds}{dt}\right)^3}{Y\left(\frac{dx}{dt}\right)} = \frac{\left(\frac{ds}{dt}\right)^3}{\frac{dx}{dt}\frac{d^2y}{dt^2} - \frac{dy}{dt}\frac{d^2x}{dt^2}} = \rho,$$

ρ being the radius of curvature. Hence

$$\left(\frac{ds}{dt}\right)^2 = Y\rho\frac{dx}{ds}$$

$$= 2Y\left(\tfrac{1}{4}\text{ chord of curvature parallel to } y\right),$$

and therefore the velocity at any point is the same as would be acquired by a particle falling in vacuo through one fourth of the chord of curvature parallel to the fixed line under the action of a uniform force equal to the force at that point.

183. PROP. *A particle moving in a resisting medium is acted on by a central force; to find the resistance that a given curve may be described.*

P denoting the central force and R the resistance, the equations of motion are

$$\frac{d^2x}{dt^2} = -P\frac{x}{r} - R\frac{dy}{ds},$$

$$\frac{d^2y}{dt^2} = -P\frac{y}{r} - R\frac{dy}{ds}.$$

Eliminating R,

$$\frac{dx}{dt}\frac{d^2y}{dt^2} - \frac{dy}{dt}\frac{d^2x}{dt^2} = \frac{P}{r}\left(x\frac{dy}{dt} - y\frac{dx}{dt}\right),$$

$$\text{or } \frac{1}{\rho}\left(\frac{ds}{dt}\right)^3 = P\frac{p}{r}\frac{ds}{dt},$$

$$\therefore \left(\frac{ds}{dt}\right)^2 = Pp\,\frac{dr}{dp} \quad\ldots\ldots\ldots\ldots\ldots\ldots\ldots\ldots \quad (1),$$

which shews that the velocity is that due to one fourth the chord of curvature through the center.

Again, eliminating P from the equations of motion, we obtain

$$x\frac{d^2y}{dt^2} - y\frac{d^2x}{dt^2} = -R\left(x\frac{dy}{dt} - y\frac{dx}{dt}\right)\frac{dt}{ds},$$

or

$$\frac{d}{dt}\left(p\frac{ds}{dt}\right) = -Rp;$$

$$\therefore p\frac{ds}{dt}\frac{d}{ds}\left(p\frac{ds}{dt}\right) = -Rp^2;$$

$$\therefore -2Rp^2 = \frac{d}{ds}\left\{p^2\left(\frac{ds}{dt}\right)^2\right\}$$

$$= \frac{d}{ds}\left(Pp^3\frac{dr}{dp}\right), \text{ by (1)};$$

$$\therefore R = -\frac{1}{2p^2}\frac{d}{ds}\left(Pp^3\frac{dr}{dp}\right) \quad\ldots\ldots \quad (2),$$

which determines the required law of resistance.

184. If the resistance vary as the product of the density and the nth power of the velocity,

$$R = k\left(\frac{ds}{dt}\right)^n,$$

and by equations (1) and (2),

$$k\left(Pp\frac{dr}{dp}\right)^{\frac{n}{2}} = -\frac{1}{2p^2}\frac{d}{ds}\left(Pp^3\frac{dr}{dp}\right),$$

$$\therefore k = -\frac{p^{n-2}}{2}\left(Pp^3\frac{dr}{dp}\right)^{-\frac{n}{2}}\frac{d}{ds}\left(Pp^3\frac{dr}{dp}\right),$$

which according as n is not or is equal to 2, may be written

$$k = \frac{p^{n-2}}{n-2} \frac{d}{ds} \left(Pp^3 \frac{dr}{dp}\right)^{\frac{2-n}{2}},$$

$$\text{or } k = -\frac{1}{2} \frac{d}{ds} \log \left(Pp^3 \frac{dr}{dp}\right),$$

which equations determine either the law of density that a given curve may be described, the force being supposed known, or the law of force supposing the density known.

185. PROP. *A particle moving in a resisting medium of which the resistance varies as the product of the density and square of the velocity is acted on by a central force; to determine the orbit.*

This may be derived immediately from the result of last Article, but we will here give a direct investigation. Let P be the central force; then taking the equations of motion along the radius vector, and perpendicular to it, we have

$$\frac{d^2r}{dt^2} - r\left(\frac{d\theta}{dt}\right)^2 = -P - k\left(\frac{ds}{dt}\right)^2 \frac{dr}{ds} \quad\text{...........(1)},$$

$$\frac{1}{r} \frac{d}{dt}\left(r^2 \frac{d\theta}{dt}\right) = -k\left(\frac{ds}{dt}\right)^2 \frac{rd\theta}{ds}.$$

Dividing the latter of these equations by $r\dfrac{d\theta}{dt}$, we have

$$\frac{\frac{d}{dt}\left(r^2 \frac{d\theta}{dt}\right)}{r^2 \frac{d\theta}{dt}} = -k\frac{ds}{dt}.$$

Integrating

$$r^2 \frac{d\theta}{dt} = h\epsilon^{-\int k ds},$$

h being the constant introduced in the integration and depending on the initial circumstances.

Again, putting $r = \frac{1}{u}$;

$$\frac{dr}{dt} = -\frac{1}{u^2}\frac{du}{dt} = -\frac{1}{u^2}\frac{du}{d\theta}\frac{d\theta}{dt}$$

$$= -h\frac{du}{d\theta}\epsilon^{-\int k ds};$$

$$\frac{d^2r}{dt^2} = -h\frac{d^2u}{d\theta^2}\epsilon^{-\int k ds}\frac{d\theta}{dt} + hk\frac{du}{d\theta}\epsilon^{-\int k ds}\frac{ds}{dt}$$

$$= -h^2u^2\frac{d^2u}{d\theta^2}\epsilon^{-2\int k ds} - k\frac{ds}{dt}\frac{dr}{dt}.$$

Substituting in equation (1), we have, since

$$\left(\frac{ds}{dt}\right)^2\frac{dr}{ds} = \frac{ds}{dt}\frac{dr}{dt}$$

$$-h^2u^2\frac{d^2u}{d\theta^2}\epsilon^{-2\int k ds} - h^2u^3\epsilon^{-2\int k ds} = -P;$$

$$\therefore \frac{d^2u}{d\theta^2} + u - \frac{P\epsilon^{2\int k ds}}{h^2u^2} = 0.$$

EXAMPLES.

(1) A particle is projected with a given velocity V in a uniform medium in which the resistance varies as the square root of the velocity; to find what time will elapse before the particle is reduced to rest.

Ans. Required time $= \frac{2V^{\frac{1}{2}}}{k}$.

(2) A particle projected with a velocity of 1000 feet a second, loses half its velocity by passing through 3 inches of a resisting medium of which the resistance is uniform; to find the time of passing through this space.

Ans. $\frac{1}{3000}$th of a second.

(3) A particle falls towards a center of force which varies as the inverse cube of the distance, in a medium of which the density varies also as the inverse cube, and of which the resistance varies as the square of the velocity; prove that at any distance x from the center,

$$(\text{velocity})^2 = \frac{\mu}{k}\left\{1 - \epsilon^{-k\left(\frac{1}{x^2} - \frac{1}{a^2}\right)}\right\},$$

where μ = force at unit of distance, k = density at unit of distance, and a = distance of the particle from the center at the beginning of its motion.

The equation of motion is

$$\frac{d^2x}{dt^2} = -\frac{\mu}{x^3} + \frac{k}{x^3}\left(\frac{dx}{dt}\right)^2;$$

$$\text{or } v\frac{dv}{dx} = -\frac{\mu}{x^3} + \frac{kv^2}{x^3}.$$

Let $v^2 = z$;

$$\therefore \frac{dz}{dx} - \frac{2kz}{x^3} = -\frac{2\mu}{x^3}.$$

Multiply by $\epsilon^{\frac{k}{x^2}}$ and integrate;

$$\therefore \epsilon^{\frac{k}{x^2}} z = C + \frac{\mu}{k}\epsilon^{\frac{k}{x^2}}.$$

Now when $x = a$, $z = v^2 = 0$; $\therefore C = -\frac{\mu}{k}\epsilon^{\frac{k}{a^2}}$;

$$\therefore z = v^2 = \frac{\mu}{k}\left\{1 - \epsilon^{-k\left(\frac{1}{x^2} - \frac{1}{a^2}\right)}\right\}.$$

(4) A particle acted on by a constant force f, moves from rest in a medium of which the resistance varies as the square of the velocity directly and as the distance from the center inversely; find the velocity of the particle at any distance

14—2

from the origin, and the position of the particle when its velocity is a maximum.

$$\text{Ans.} \quad v^2 = \frac{2f}{1-2k}\left(a^{1-2k} - x^{1-2k}\right), \quad x' = (2k)^{\frac{1}{1-2k}}\, a,$$

where a is the initial distance and k the resistance when x and v are both unity.

(5) If chords be drawn from either extremity of a vertical diameter of a circle, the time of descent down each of them in a medium whose resistance varies as the velocity, is the same.

(6) A particle is projected with a given velocity, towards a center of force attracting inversely as the cube of the distance, in a medium of which the resistance varies as the square of the velocity directly, and as the square of the distance from the center inversely; to find the velocity at any point.

$$\text{Ans.} \quad \epsilon^{-\frac{2k}{a}} v^2 - \epsilon^{-\frac{2k}{x}} V^2 = \frac{\mu}{2k^2}\left(\frac{a-2k}{a}\epsilon^{-\frac{2k}{x}} - \frac{x-2k}{x}\epsilon^{-\frac{2k}{a}}\right),$$

where V is the velocity of projection, μ the absolute attracting force, k the resistance at a unit of distance, a the initial distance, and x the distance of the particle corresponding to the velocity v.

(7) One particle begins to fall from the higher extremity of a vertical line, and at the same instant another is projected upwards from the other extremity with a given velocity, the particles moving in a medium of which the resistance varies directly as the velocity; shew that the time at which they will meet $= \frac{1}{k}\log\frac{V}{V-ka}$ where a is the length of the vertical line, V the velocity of projection, and k the resistance for a unit of velocity.

(8) A particle acted on by gravity falls from a given altitude in an uniform medium of which the resistance varies as the square of the velocity; on arriving at the lowest point of its descent it is reflected upwards with the velocity which it has acquired in its fall, after reaching its greatest altitude it again descends and is again reflected, and so on perpetually; to determine the altitude of ascent after any number of reflections.

Let a be the height from which the particle falls, a_1, a_2, a_3, &c., its maximum altitudes afterwards, v_1, v_2, v_3, &c., its velocities after the first, second, &c. times of reaching the bottom, then (Art. 173),

$$v_1^2 = K^2 (1 - \epsilon^{-\frac{2fa}{K^2}}),$$

and

$$\frac{1}{v_2^2} - \frac{1}{v_1^2} = \frac{1}{K^2},$$

$$\frac{1}{v_3^2} - \frac{1}{v_2^2} = \frac{1}{K^2},$$

$$\cdots\cdots\cdots\cdots\cdots\cdots$$

$$\cdots\cdots\cdots\cdots\cdots\cdots$$

$$\frac{1}{v_n^2} - \frac{1}{v_{n-1}^2} = \frac{1}{K^2}.$$

Adding

$$\frac{1}{v_n^2} - \frac{1}{v_1^2} = \frac{n-1}{K^2}.$$

Hence

$$\frac{1}{v_n^2} = \frac{1}{K^2}\left(n - 1 + \frac{1}{1 - \epsilon^{-\frac{2fa}{K^2}}}\right)$$

$$= \frac{1}{K^2}\left\{\frac{n - (n-1)\,\epsilon^{-\frac{2fa}{K^2}}}{1 - \epsilon^{-\frac{2fa}{K^2}}}\right\};$$

$$\therefore \frac{v_n^2}{K^2} = \frac{\epsilon^{\frac{2fa}{K^2}} - 1}{n\epsilon^{\frac{2fa}{K^2}} - n + 1}.$$

The particle is now projected upwards with a velocity v_n, therefore (Art. 173),

$$a_n = \frac{K^2}{2f}\log\left(1 + \frac{v_n^2}{K^2}\right)$$

$$= \frac{K^2}{2f}\log\frac{(n+1)\,\epsilon^{\frac{2fa}{K^2}} - n}{n\epsilon^{\frac{2fa}{K^2}} - n + 1}.$$

If a be equal to infinity,

$$a_n = \frac{K^2}{2f} \log \frac{n+1}{n}.$$

(9) To determine the laws of force that a particle may always descend to a given center in the same time from whatever distance it commences its motion, the medium in which the particle moves being uniform, and the resistance varying as the square of the velocity.

(10) If a particle be projected in a medium, the resistance of which varies as the velocity, and be acted on by gravity, and another projected in vacuo at the same angle and with the same velocity and acted on by the same constant force, and if t_1, t_2 be the times of describing two arcs in the medium and in vacuo so related to each other that the tangents at their extremities shall cut the axis of x at the same angle, then $\epsilon^{kt_1} - 1 = kt_2$, k being the resistance corresponding to the velocity 1.

(11) A particle moves in a semicircle, acted on by a constant force in parallel lines; find the requisite resistance, and supposing the resistance to vary as the density into the square of the velocity, find the law of density.

Let the equation to the circle be $x^2 + y^2 = a^2$, and the force parallel to y; then $R \propto x$, $k \propto \dfrac{x}{(a^2 - x^2)^{\frac{1}{2}}}$.

(12) A particle acted on by a constant force parallel to the axis of y moves in the curve $a^{n-1}y + x^n = b^n$; find the law of resistance.

$$\text{Ans.} \quad R \propto \frac{\sqrt{(a^{2n-2} + n^2 x^{2n-2})}}{x^{n-1}}.$$

(13) A particle acted on by a constant force parallel to the axis of y moves in the curve, $y = mx + \dfrac{a^{n+1}}{x^n}$; find the law of density, the resistance varying as the product of the density and square of the velocity.

$$\text{Ans.} \quad k \propto \frac{x^n}{\{x^{2n+2} + (mx^{n+1} - na^{n+1})^2\}^{\frac{1}{2}}}.$$

(14) A particle moves in a circle about a center of force in the circumference, the force being attractive and $=\mu r^n$; to find the resistance of the medium and the law of the density. Supposing the resistance equal to the product of the density and the square of the velocity.

$$\text{Ans.} \quad R = \frac{\mu}{4}(n+5)\, r^n \sin\theta,\ k = \frac{n+5}{2}\,\frac{\sin\theta}{r}.$$

(15) A particle moves towards the pole in an equiangular spiral about a center of force in the pole the force being μr^n; to find the resistance and density of the medium, the resistance being equal to the product of the density and the square of the velocity.

$$\text{Ans.} \quad R = \frac{\mu(n+3)}{2}\, r^n \cos\alpha,\ k = \frac{n+3}{2}\,\frac{\cos\alpha}{r},$$

where α is the constant angle of the spiral.

(16) A particle moves in the circumference of a circle about a center of force in the center, the resistance of the medium in which the motion takes place being constant; to find the law of force, the velocity at any time, and the time which elapses as well as the space described before the particle is reduced to rest.

$$\text{Ans.} \quad v^2 = V^2 - 2cs,\ P = \frac{1}{a}(V^2 - 2cs),\ S = \frac{V^2}{2c},\ T = \frac{V}{C},$$

where $V =$ initial velocity, $c =$ constant force of resistance, $a =$ radius of the circle, s the arc described from the beginning of the motion, S and T the space and time corresponding to the particle's being reduced to rest.

(17) A particle is projected along a smooth curve with velocity V in a medium whose resistance $= kv^2$. Prove that when the direction of the motion has changed through an angle ϕ the velocity $= V\epsilon^{-k\phi}$.

(18) A particle moves towards the pole in an equiangular spiral, the motion taking place in a medium whose resistance $= kr^n$; find the law of central attractive force in the pole.

$$\text{Ans.} \quad P = \frac{(n+3)\cos\alpha\, a^2 V^2 + 2k\,(r^{n+3} - a^{n+3})}{(n+3)\cos\alpha\, r^3},$$

where $V =$ initial velocity and $\alpha =$ constant angle of the spiral.

(19) If a particle acted on by a central force P is moving in a medium whose resistance $= k$ (velocity) prove that

$$\frac{d^2r}{dt^2} + P - \frac{h^2}{r^3}\epsilon^{-2kt} + k\frac{dr}{dt} = 0,$$

where h is an arbitrary constant.

(20) A heavy particle moving in a medium whose resistance $= nv^2$, is compelled to describe in a vertical plane the curve

$$ax = \epsilon^{ns} - ns - 1,$$

where s is the length of the curve measured from the lowest point, x the abscissa of the extremity of this arc referred to a vertical axis, and a a constant; shew that the time of reaching the lowest point is independent of the height from which it starts.

(21) Shew that the curve in last question is also tautochronous if the resistance $= mv + nv^2$.

CHAPTER IX.

GENERAL THEOREMS.

186. WE propose now to prove some of the general theorems connected with the motion of a particle under the action of any forces, and to investigate the forces requisite for the description of given paths in a given manner.

187. PROP. *If a particle be subject to the action of forces, whose resultant is continually at right angles to the direction of its motion; the velocity will be uniform.*

Let R be this resultant λ, μ, ν, its direction cosines, then if m be the mass of the particle,

$$m\frac{d^2x}{dt^2} = \lambda R,$$

$$m\frac{d^2y}{dt^2} = \mu R,$$

$$m\frac{d^2z}{dt^2} = \nu R.$$

Multiplying by $\frac{dx}{dt}$ ... adding, and observing that

$$\lambda\frac{dx}{ds} + \mu\frac{dy}{ds} + \nu\frac{dz}{ds} = 0,$$

since the force R is at right angles to the element of the path,

$$m\frac{d}{dt}(v^2) = 2m\left(\frac{dx}{dt}\frac{d^2x}{dt^2} + \frac{dy}{dt}\frac{d^2y}{dt^2} + \frac{dz}{dt}\frac{d^2z}{dt^2}\right) = 0;$$

or $v = \text{const.}$

Or, we might at once have resolved along the arc; this would have given

$$\frac{d^2s}{dt^2} = 0;$$

$$\text{or integrating, } \frac{ds}{dt} = v = \text{const.}$$

The value of R (Art. 10) is evidently $\frac{mv^2}{\rho}$; or varies inversely as the radius of absolute curvature of the path; and it is clear that it lies in the osculating plane, since there is no acceleration perpendicular to that plane.

188. Ex. I. Suppose a particle projected in a plane is acted on by a constant force R in that plane continually perpendicular to its direction of motion; find the path described.

Here $R = \frac{mv^2}{\rho}$; and $\therefore \rho$ is constant, or the path is a circle.

Ex. II. If R vary as the time elapsed since the commencement of the motion $R = R_0 t$.

Also s the arc of the curve described in the same time $= vt$, since v is constant;

$$\therefore R = \frac{mv^2}{\rho} = R_0 \frac{s}{v};$$

$$\therefore \rho s = \text{const.} = c^2, \text{ suppose.}$$

If ψ be the angle which the direction of the element δs makes with the axis of x, $\rho = \frac{ds}{d\psi}$;

$$\therefore s \frac{ds}{d\psi} = c^2,$$

$$s^2 = 2c^2 (\psi + C),$$

the equation to the curve described.

189. Prop. *If X, Y, Z, be the resolved parts of a force or forces such as occur in nature, i. e. tending to fixed centers and being functions of the distances from these centers,*

$$Xdx + Ydy + Zdz,$$

is a perfect differential.

Let the points a_1, b_1, c_1; a_2, b_2, c_2; &c., be the positions of the centers of force; x, y, z the co-ordinates of the particle; then if $r_1, r_2, \ldots$ be its distances from the centers,

$$\phi_1'(D),\ \phi_2'(D),\ \&c.,$$

the laws of attraction to those centers; we have

$$X = \frac{a_1 - x}{r_1}\phi_1'(r_1) + \frac{a_2 - x}{r_2}\phi_2'(r_2) + \ldots\ldots\ldots$$

$$= \Sigma \frac{a - x}{r}\phi'(r).$$

But $r = \sqrt{\{(a-x)^2 + (b-y)^2 + (c-z)^2\}}$;

$$\therefore \left(\frac{dr}{dx}\right) = -\frac{a-x}{r},$$

$$\text{or } X = -\Sigma\phi'(r)\left(\frac{dr}{dx}\right),$$

$$Y = -\Sigma\phi'(r)\left(\frac{dr}{dy}\right),$$

$$Z = -\Sigma\phi'(r)\left(\frac{dr}{dz}\right),$$

where the differential coefficients are partial;

$$\therefore Xdx + Ydy + Zdz$$

$$= -\Sigma\phi'(r)\left\{\left(\frac{dr}{dx}\right)dx + \left(\frac{dr}{dy}\right)dy + \left(\frac{dr}{dz}\right)dz\right\}$$

$$= -\Sigma\phi'(r)\,dr \ldots\ldots\ldots\ldots\ldots\ldots\ldots\ldots\ldots(1).$$

And thus every term of this sum is a perfect differential, which proves the proposition.

190. PROP. *Under the action of any forces such as occur in nature the increment of the square of the velocity of a particle in passing from one point to another is independent of the path*

pursued, and depends only on the initial and final positions. This is true even if the particle be constrained to move in any particular path by a force continually perpendicular to its direction of motion.

Taking tangential resolution, the force of constraint disappears, and

$$\frac{d^2s}{dt^2} = S.$$

Now S, the resolved accelerating force along the tangent, is

$$X\frac{dx}{ds} + Y\frac{dy}{ds} + Z\frac{dz}{ds},$$

and therefore by (1),

$$\frac{ds}{dt}\frac{d^2s}{dt^2} = -\Sigma\phi'(r)\frac{dr}{dt},$$

$$\text{or } v^2 = C - 2\Sigma\phi(r);$$

∴ if V be the velocity at a point whose distances from the centers are R_1, R_2............

$$v^2 - V^2 = 2\{\Sigma\phi(R) - \Sigma\phi(r)\},$$

a result which involves only the co-ordinates of the initial and final positions.

191. Hence if from any point of the surface $\Sigma\phi(r) = A$, a particle be projected with a given velocity in any direction; its velocity will be the same in whatever point it meet the surface $\Sigma\phi(r) = B$, A and B being any constants.

Now on account of equation (1), $\Sigma\phi(r) =$ constant is the equation to a surface on which if smooth a particle will rest in any position under the action of the given forces.

Hence a particle leaving any point of a surface of equilibrium with a given velocity, will have on reaching any other surface of equilibrium a velocity independent of the path pursued or the point reached.

DEF. *The product of the mass, and the square of the velocity, of a particle is called the Vis Viva.*

The last proposition contains the principle of Vis Viva for the case of a single particle.

From this it follows that if several particles moving under the action of the same center of force have equal velocities at any particular distance from the center; their velocities will always be equal at equal distances from that center.

Now we have seen (Art. 100) that the axis major $2a$ of an elliptic orbit about a center of force in the focus is independent of the direction of projection. Hence, by considering the projection to be *from* the center, we find that the velocity at any point is that due to a fall from rest at a distance $2a$, to that point; and that therefore in any elliptic orbit about a focus the velocity at any point is that due to a fall to that point, through a space equal to the distance from the other focus.

192. PROP. *If a number of separate particles whose masses are* m_1, m_2, &c. *subjected to the action of forces* f_1, f_2, *respectively (these forces being such as occur in nature), and projected from the same point in the same direction with velocities* $\overline{v}_1$, $\overline{v}_2$, &c., *describe the same path; the latter will also be described by a mass* M *projected with velocity* $\overline{V}$ *from the same point in the same direction, and acted on at once by the forces* f_1, f_2, &c. *provided*

$$M\overline{V}^2 = \Sigma (m\overline{v}^2).$$

Suppose that in addition to the forces f_1, f_2, &c., a force R continually acting in a direction at right angles to that of M's motion be required to cause it to move in the given path; the equations of motion are

$$M\frac{d^2x}{dt^2} = M\Sigma(X) + R\lambda \;\ldots\ldots\ldots\ldots\ldots\; (1),$$

with similar equations in y and z,

where λ, μ, ν are the direction cosines of R, and X, Y, Z the resolved parts of f.

Multiplying by $\frac{dx}{ds}$, $\frac{dy}{ds}$, $\frac{dz}{ds}$ in order, and adding, R goes out, and we have

$$Md(V^2) = 2\{\Sigma(X)dx + \Sigma(Y)dy + \Sigma(Z)dz\}.$$

But for the separate particles m_1, m_2, &c. we have

$$m_1 d(v_1^2) = 2(X_1 dx + Y_1 dy + Z_1 dz), \text{ \&c.}$$

therefore

$$\Sigma\{md(v^2)\} = 2\{\Sigma(X)dx + \Sigma(Y)dy + \Sigma(Z)dz\};$$

$$\therefore \Sigma\{md(v^2)\} = Md(V^2),$$

$$\text{or } \Sigma(mv^2) = MV^2 + C.$$

But $\Sigma(m\bar{v}^2) = M\bar{V}^2$ by hypothesis, therefore $C = 0$.

Hence the vis viva of M will be at each point of the orbit equal to the sum of the vires vivæ of m_1, m_2, &c., at that point. To find R notice that in general the pressure on a constraining curve is the sum of the resolved parts of the impressed forces, and the pressure due to the velocity. Now the latter part is as the vis viva, therefore, in the case of M it is the sum of the same forces in the case of m_1, m_2, &c. Also the same may be said of the resolved parts of the impressed forces. But in the case of each particle, these partial pressures destroyed each other since the curve was described freely, hence their sums will destroy each other, or the curve will be freely described by M.

193. Def. If v be the velocity of a particle whose mass is m, and s be the length of the path described, the value of the integral $m\int vds$ taken between proper limits is called the Action of the particle.

Prop. *If a particle move freely, or on a smooth surface, (under the action of forces such as occur in nature,) between any two points, the value of the integral $m\int vds$ for the whole actual path is generally less than it would be if the particle were constrained to pass from one point to the other by a different path.* (This, combined with the above definition, is the Principle of Least Action).

If δ be the symbol of the Calculus of Variations, the proposition will be proved if we shew that

$$\delta\int v\,ds = 0.$$

Now $\delta\int v\,ds = \int\delta(v\,ds) = \int(v\delta ds + ds\,\delta v)$

$$= \int(v\delta ds + \tfrac{1}{2}dt\,\delta v^2) \text{ since } v = \frac{ds}{dt}.$$

But generally $v^2 = 2\int(Xdx + Ydy + Zdz) = \psi(x, y, z)$, the force of constraint having disappeared;

$$\therefore\ \tfrac{1}{2}\delta v^2 = X\delta x + Y\delta y + Z\delta z.$$

But (Art. 148) $X = \dfrac{d^2x}{dt^2} - \dfrac{R}{m}\lambda.$

&c.

Hence

$$\tfrac{1}{2}\delta v^2 = \left(\frac{d^2x}{dt^2}\delta x + \frac{d^2y}{dt^2}\delta y + \frac{d^2z}{dt^2}\delta z\right)$$

$$-\frac{R}{m}(\lambda\delta x + \mu\delta y + \nu\delta z).$$

Now if the particle remain on the surface whose equation is $F = 0$,

$$\lambda\delta x + \mu\delta y + \nu\delta z = \delta F = 0,$$

and if it leave it $R = 0$, so in either case the latter portion vanishes.

$$\text{Also } ds^2 = dx^2 + dy^2 + dz^2;$$

$$\therefore\ ds\delta ds = dx\,\delta dx + dy\,\delta dy + dz\,\delta dz,$$

$$\text{or } v\delta ds = \frac{dx}{dt}\delta dx + \frac{dy}{dt}\delta dy + \frac{dz}{dt}\delta dz$$

$$= \frac{dx}{dt}d\delta x + \frac{dy}{dt}d\delta y + \frac{dz}{dt}d\delta z.$$

since the order of d and δ is immaterial.

Hence, $\delta\int v ds = \int\left\{\frac{dx}{dt}\, d\delta x + \frac{dy}{dt}\, d\delta y + \frac{dz}{dt}\, d\delta z\right.$

$$\left. + \delta x d\left(\frac{dx}{dt}\right) + \delta y d\left(\frac{dy}{dt}\right) + \delta z d\left(\frac{dz}{dt}\right)\right\}$$

$$= \frac{dx}{dt}\,\delta x + \frac{dy}{dt}\,\delta y + \frac{dz}{dt}\,\delta z$$

taken between proper limits. Now at both limits

$$\delta x = 0, \quad \delta y = 0, \quad \delta z = 0;$$

$$\therefore \text{ we have } \delta\int v ds = 0.$$

And, as generally it is impossible to suppose the action a maximum, this equation shews that it is a minimum.

If no forces act on the particle except the constraint of the surface, we have v constant, and the above equation shews that in this case $\int ds$ or the length of the path is a minimum.

A particle therefore, projected along a smooth surface and subject to no forces, will trace out between any two points in its path the shortest line on the surface. (Art. 150).

It may happen in the case of a closed surface, a sphere for instance, that the particle will trace out the longest line on the surface between the two points; in this case the value of the action is a maximum.

194. We may apply this principle directly to form the equations of motion in any particular case, or to find the actual path under the action of any forces.

Ex. (1). Let us take again the case of the refraction of light in the corpuscular theory. (Art. 80).

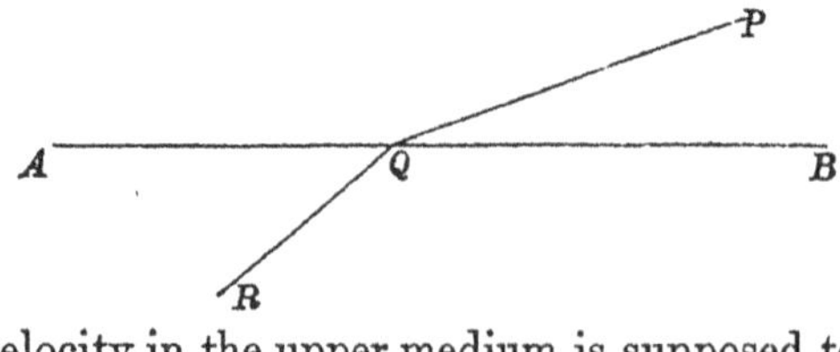

The velocity in the upper medium is supposed to be u, that in the lower v.

In this case the expression for the action becomes simply

$$uPQ + vPR,$$

if PQR be the path of the particle.

By making this quantity a minimum, as depending on the position of Q, it is easy to find that Q must lie in the plane through P and R perpendicular to the surface AB, and also that the sines of the inclinations of PQ and QR to the normal at Q must be in the inverse ratio of the velocities.

195. Ex 2. To find the equation to the path described by a particle about a center of force.

Let P be the central force at distance r, then

$$v^2 = -2\int Pdr, \text{ (Art. 92)}$$

$$= \{\phi(r)\}^2, \text{ suppose, } \ldots\ldots\ldots\ldots\ldots \text{(1)},$$

$$\therefore \int vds = \int \phi(r)\, ds,$$

or, making x the independent variable,

$$= \int \phi(r) \frac{ds}{dx} dx = \int \phi(r) \sqrt{\left(1 + \overline{\frac{dy}{dx}}\Big|^2 + \overline{\frac{dz}{dx}}\Big|^2\right)} \cdot dx.$$

This is to be a minimum by the variation of the expressions for y and z in terms of x, and

$$\therefore \frac{d}{dx}\left\{\frac{\frac{dy}{dx}}{\phi(r)\sqrt{\left(1 + \overline{\frac{dy}{dx}}\Big|^2 + \overline{\frac{dz}{dx}}\Big|^2\right)}}\right\} = 0,$$

$$\frac{d}{dx}\left\{\frac{\frac{dz}{dx}}{\phi(r)\sqrt{\left(1 + \overline{\frac{dy}{dx}}\Big|^2 + \overline{\frac{dz}{dx}}\Big|^2\right)}}\right\} = 0.$$

Integrating and dividing one by the other,

$$\frac{\frac{dy}{dx}}{\frac{dz}{dx}} = \text{constant}.$$

Hence, the curve is plane, and it will be preferable to refer it to polar co-ordinates in its plane.

We have therefore

$$\int \phi(r) \sqrt{\left\{r^2 + \left(\frac{dr}{d\theta}\right)^2\right\}} \cdot d\theta, \text{ a minimum,}$$

and by the formula $V = Pp + C$,

$$\phi(r) \sqrt{\left\{r^2 + \left(\frac{dr}{d\theta}\right)^2\right\}} = \phi(r) \frac{\left(\frac{dr}{d\theta}\right)^2}{\sqrt{\left\{r^2 + \left(\frac{dr}{d\theta}\right)^2\right\}}} + C,$$

or reducing, and putting h for C,

$$r^2 \phi(r) = h \sqrt{\left\{r^2 + \left(\frac{dr}{d\theta}\right)^2\right\}} \quad \cdots\cdots\cdots (2),$$

$$\text{or } r^2 \frac{ds}{dt} = h \frac{ds}{d\theta};$$

$$\therefore r^2 \frac{d\theta}{dt} = h,$$

the equation for the equable description of areas.

Squaring (2) and attending to (1), we have,

$$-\frac{2r^4}{h^2} \int P dr = r^2 + \left(\frac{dr}{d\theta}\right)^2,$$

$$\text{or putting } r = \frac{1}{u},$$

$$\frac{2}{h^2} \int \frac{P}{u^2} \frac{du}{d\theta} d\theta = u^2 + \left(\frac{du}{d\theta}\right)^2,$$

or, differentiating and dividing by $2\frac{du}{d\theta}$,

$$\frac{P}{h^2 u^2} = u + \frac{d^2 u}{d\theta^2},$$

the general equation to central orbits.

196. When a particle moves in any curve, it has been shewn (Art. 10), that the acceleration along the radius of absolute curvature of the path is $\frac{v^2}{\rho}$; that is, a force $\frac{mv^2}{\rho}$ is required to deflect the particle from the tangent, which is the path it would take if left to itself.

From this, or by the formula in Art. 93, we see that if a particle revolve at distance r, with angular velocity ω, about a point, a force $mr\omega^2$ to that point is requisite to maintain the distance r unaltered. This tendency to move in the tangent, which arises from the inertia of matter, was formerly supposed to be due to a force called *Centrifugal Force* generated in the particle by its rotation about the point.

We have seen that when the motion of a particle in any path is referred to polar co-ordinates in a plane, the acceleration along the radius vector is

$$\frac{d^2r}{dt^2} - r\left(\frac{d\theta}{dt}\right)^2.$$

Now the velocity along r is $\frac{dr}{dt}$, and that perpendicular to it $r\frac{d\theta}{dt}$; hence the first term of the above is the acceleration of the velocity along the radius vector, and the other is the so-called centrifugal force due to a velocity $r\frac{d\theta}{dt}$ in a circle of radius r. The idea of this force is useful as we have already seen (Art. 164), in enabling us to form the equations of motion of a particle in particular cases.

197. PROP. *The motion of a particle in a plane being given with respect to fixed axes, to investigate expressions for the velocity and acceleration parallel to axes about the same origin, which revolve in the plane with uniform angular velocity.*

Let the angular velocity be ω, then if at time $t=0$, the two sets of axes be coincident, at time t they will include an angle ωt; and therefore x, y, x', y', being at that instant the co-ordinates of the particle with respect to the two sets of axes,

$$x' = x \cos \omega t + y \sin \omega t,$$

$$y' = y \cos \omega t - x \sin \omega t;$$

$$\therefore \frac{dx'}{dt} = \frac{dx}{dt} \cos \omega t + \frac{dy}{dt} \sin \omega t - \omega (x \sin \omega t - y \cos \omega t)$$

$$= \frac{dx}{dt} \cos \omega t + \frac{dy}{dt} \sin \omega t + \omega y'.$$

Similarly, $\frac{dy'}{dt} = \frac{dy}{dt} \cos \omega t - \frac{dx}{dt} \sin \omega t - \omega x'$,

which determine the velocities.

Again,

$$\frac{d^2x'}{dt^2} = \frac{d^2x}{dt^2} \cos \omega t + \frac{d^2y}{dt^2} \sin \omega t$$

$$+ 2\omega \frac{dy}{dt} \cos \omega t - 2\omega \frac{dx}{dt} \sin \omega t - \omega^2 x',$$

and $\frac{d^2y'}{dt^2} = \frac{d^2y}{dt^2} \cos \omega t - \frac{d^2x}{dt^2} \sin \omega t$

$$- 2\omega \frac{dx}{dt} \cos \omega t - 2\omega \frac{dy}{dt} \sin \omega t - \omega^2 y',$$

the required accelerations.

198. PROP. *Given the path of a particle, and the manner of its description, to find the requisite forces.*

If mX, mY, mZ be the required forces, we must have

$$X = \frac{d^2x}{dt^2} = \frac{d}{dt}\left(\frac{dx}{dt}\right)$$

$$= \frac{ds}{dt} \frac{d}{ds}\left(\frac{ds}{dt} \frac{dx}{ds}\right) = v \frac{d}{ds}\left(v \frac{dx}{ds}\right),$$

with similar expressions for Y and Z. But as the path is given, and the manner of its description, that is v in terms of the co-ordinates, the value of the above expressions is completely known.

Instead of having the velocity at any point, we might have had other conditions, such for instance as that the resultant force is to be in a given direction, &c., but these like the above present no difficulty.

199. Def. *If the radius vector of a curve in space be at each instant parallel to the direction, and equal to the magnitude, of the velocity of a particle moving in any path; the curve is called the hodograph corresponding to the path.*

The hodograph is evidently a plane curve if the path is so.

Let x, y, z be the co-ordinates of a point in the path, x', y', z' those of the corresponding point of the hodograph; then evidently by the Definition,

$$\left.\begin{aligned}\frac{dx}{dt} &= x'\\ \frac{dy}{dt} &= y'\\ \frac{dz}{dt} &= z'\end{aligned}\right\}.$$

Hence if s' be the arc of the hodograph

$$\frac{ds'}{dt} = \sqrt{\left\{\left(\frac{dx'}{dt}\right)^2 + \left(\frac{dy'}{dt}\right)^2 + \left(\frac{dz'}{dt}\right)^2\right\}}$$
$$= \sqrt{\left\{\left(\frac{d^2x}{dt^2}\right)^2 + \left(\frac{d^2y}{dt^2}\right)^2 + \left(\frac{d^2z}{dt^2}\right)^2\right\}},$$

and the direction cosines of $\delta s'$ are proportional to

$$\frac{d^2x}{dt^2},\ \frac{d^2y}{dt^2},\ \frac{d^2z}{dt^2}.$$

Hence we see that

The tangent to the hodograph at any instant is parallel to the resultant force acting on the particle at the corresponding point of its path, and the elementary arc is equal to the whole acceleration of the particle during the corresponding element of time.

200. The most important case of the hodograph being that corresponding to an orbit about a single center of force, we may deduce the above properties for that case in a somewhat different manner.

Let P be any point in PA, an arc of an orbit described about a center of force S. Draw SY perpendicular to the

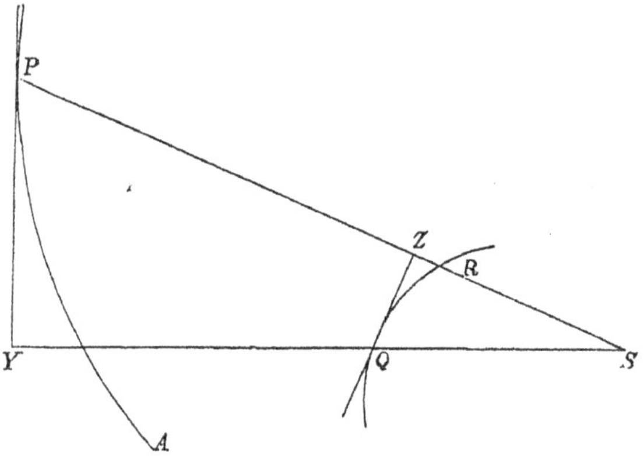

tangent at P, and take $SQ \cdot SY = h$, then evidently SQ is equal to the velocity at P, and perpendicular to it in direction. Hence the locus of Q is the hodograph turned in its own plane through 90°.

But we see that it is the polar reciprocal of PA with regard to a circle whose center is S and radius $= \sqrt{(h)}$. Hence, (SALMON, *Conics*, 308) the tangent at Q is perpendicular to SP. This evidently corresponds to the first of the two general properties of the hodograph given in last Article.

Let $r, \theta, p, s, r', \theta', p', s'$ represent the usual quantities for corresponding points of the two curves; then if ρ be the radius of curvature at Q, we have by the condition that QZ is perpendicular to SP,

$$\delta s' = \rho\delta\theta = r'\frac{dr'}{dp'}\delta\theta$$

$$= \frac{h}{p}\frac{d\frac{1}{p}}{d\frac{1}{r}}\delta\theta = \frac{h}{p^3}\frac{dp}{dr}r^2\delta\theta$$

$$= \frac{h^2}{p^3}\frac{dp}{dr}\delta t = P\delta t, \text{ (Art. 88)}$$

which proves the second property.

201. PROP. *To find the mode of description of a given hodograph that it may correspond to a central orbit.*

Here, $$\rho\delta\theta = \delta s' = \frac{r'^2\delta\theta'}{p'};$$

$$\therefore \rho r^2\frac{d\theta}{dt} = \frac{h^2 r'^2}{p'^3}\cdot\frac{d\theta'}{dt},$$

$$\text{or } h\frac{d\theta'}{dt} = \frac{\rho p'^3}{r'^2} = \frac{\rho r'^4}{\left(\frac{ds'}{d\theta'}\right)^3},$$

which gives the required angular velocity at any point of the hodograph, in terms of the co-ordinates of that point.

202. When the central force is inversely as the square of the distance, we have by Art. 200 for the arc of the hodograph,

$$\delta s' = \frac{\mu}{r^2}\delta t = \frac{\mu}{h}\delta\theta,$$

$$\text{or } \rho = \frac{\mu}{h}.$$

Hence for all conic sections described about the focus the hodograph is a circle.

This might have been shewn in another way, thus. In the fig. (Art. 200) if PA be a portion of an ellipse or hyperbola of which S is the focus, the locus of Y is the auxiliary circle. Hence evidently the locus of Q is a circle. If PA be a portion of a parabola of which S is the focus, the locus of Y is a straight line, and therefore that of Q is a circle passing through S.

Hence generally, the hodograph for any orbit about a center of force whose intensity is inversely as the square of the distance, is a circle; about an internal point for an ellipse, an external point for an hyperbola, and about a point in the circumference for a parabola.

203. The law of diffusion of heat and light from a calorific and luminous body is that of the inverse square of the distance. Hence an arc of the hodograph of a planet's orbit, which arc we have already seen to represent the entire acceleration due to the central force, represents also the entire amount of light or heat derived from the Sun during the passage through the corresponding arc of its orbit.

Ex. Compare the amounts of light and heat received throughout their orbits by the Earth moving in a circle, and a comet moving in a parabola at the same perihelion distance.

The hodographs are both circles, one about its center, the other about a point in its circumference; but the diameter of the latter is $\sqrt{(2)}$ times the radius of the former, (Art. 99).

Hence their circumferences are as $\sqrt{(2)} : 1$, or the Earth in its orbit receives in a revolution $\sqrt{(2)}$ times the amount of light and heat, which the comet can receive in its whole path.

204. It is evident that that diameter of the circular hodograph which passes through the center of force is divided by the center of force in the same ratio as the axis major of the orbit is divided by the focus; and by (Art. 202) its length $= \frac{2\mu}{h}$.

205. Sir W. R. Hamilton, the inventor of the hodograph, enunciates (*Quaternions*, p. 614), the following,

PROP. *If two circular hodographs, having a common chord, which passes through, or tends to, a common center of force, be both cut perpendicularly by a third circle, the times of hodographically describing the intercepted arcs will be equal.*

It is evident from (204), that the two orbits are conic sections of the same species, and with equal major axes.

Also every circle which cuts both hodographs perpendicularly must have its center on the common chord. Let the

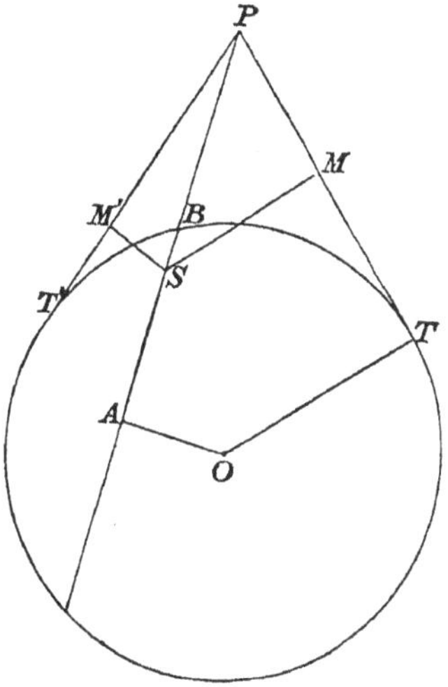

figure represent one of the hodographs, S being the center of force, and ABP the common chord. Take any point P and draw the tangents PT, PT'. We proceed to investigate the difference of the times of hodographically describing TT' and the corresponding arc for a position of P slightly shifted along AP.

Draw OA perpendicular to AP. Let $OT=a$, $AB=c$, $OA=f$, $SP=p$, $SM=\pi$, $PO=q$, $PA=p'$, and $PT=\tau$. If P be moved through a space δp, the increase of the angle PSM which is the angle vector in the orbit, is $\frac{\pi\delta p}{p\tau}$ nearly. But the corresponding radius vector in the orbit is $\frac{h}{\pi}$ and therefore the time of hodographically describing the small arc at T is

$$\delta t = \frac{h}{\pi^2}\frac{\pi\delta p}{p\tau} = \frac{\mu\delta p}{p\tau}\frac{1}{\pi a}.$$

Hence the whole change produced in the time of hodographically describing the arc TT' by shifting P is

$$\frac{\mu\delta p}{p\tau}\left(\frac{1}{a\pi}+\frac{1}{a\pi'}\right) = \frac{\mu\delta p}{p\tau}\frac{2p'}{p}\frac{q^2}{a^2p'^2-f^2\tau^2}$$

$$= \frac{2\mu p'\delta p}{p^2\tau}\frac{a^2+\tau^2}{a^2(c^2+\tau^2)-(a^2-c^2)\tau^2}$$

$$= \frac{2\mu p'\delta p}{c^2p^2\tau}.$$

Now this is the same for both hodographs, and, as the arc TT' vanishes for each when P is at B, we have the proposition.

It will readily be seen that this is in substance the same as Lambert's Theorem, (Arts. 123, 124).

206. We now take an instance of the determination, from the hodograph and the law of its description, of the curve described and the forces acting.

The hodograph is a circle described with uniform angular velocity about a point in its circumference, find the original path and the circumstances of its description.

Here we have in the hodograph,

$$\rho = a\cos\theta,$$
$$\theta = \omega t;$$

therefore in the path

$$\frac{dx}{dt} = \rho\cos\theta = a\cos^2\omega t,$$

$$\frac{dy}{dt} = \rho\sin\theta = a\cos\omega t\sin\omega t.$$

Integrating and properly adapting the constants, as they change only the origin,

$$x = \frac{a}{4\omega}(2\omega t + \sin 2\omega t),$$

$$y = \frac{a}{4\omega}(1 - \cos 2\omega t).$$

Now the equations to a cycloid are

$$x = A(\phi + \sin\phi),$$

$$y = A(1 - \cos\phi);$$

$\therefore$ the path is a cycloid; and since $2\omega t = \phi$ the direction of motion revolves uniformly. The cycloid therefore is inverted, and the particle descends from the cusp under the action of a constant force as gravity. The converse is easily proved.

EXAMPLES.

(1) INVESTIGATE the differential equation to the path of a particle in a plane

$$2X = \frac{d}{dx}\left(\frac{Y - X\frac{dy}{dx}}{\frac{d^2y}{dx^2}}\right).$$

(2) A particle slides down an inverted cycloid from rest at the cusp; shew that the whole acceleration at any instant is g, and that its direction is to the center of the generating circle.

(3) When the whole force on a particle is along the radius vector, shew that the centrifugal force from the pole

$$= \frac{h^2}{r^3}.$$

(4) A particle moves in a plane under the action of any forces, whose resolved parts are P in the radius vector, and T

in the tangent. Shew that the chord of curvature through the pole is $\dfrac{2v^2}{P+T\cot\psi}$, where ψ is the exterior angle between the radius vector and tangent.

(5) A catenary is freely described under the action of a force parallel to the axis; shew that the centrifugal force is constant.

(6) A particle, projected from the origin along the axis of y, describes the curve $y^2 = 4ax$ under the action of a force μy parallel to y, and another parallel to x; shew that

$$v^2 = \mu\left(1 + \frac{y^2}{4a^2}\right)(c^2 - y^2).$$

(7) The curve $y=\phi(x)$ touches the axis of y at the origin, and is described freely by a particle under the action of forces Y parallel to y, and f parallel to x; shew that

$$Y = 2f\sqrt{(x)}\,\frac{d}{dx}\left\{\sqrt{(x)}\,\frac{dy}{dx}\right\},$$

and that if $Y \propto x^n$, n is negative.

(8) A particle describes the hyperbola $x^2=y^2+a^2$, so that

$$\left(\frac{dx}{dt}\right)^2 = \left(\frac{dy}{dt}\right)^2 + c^2.$$

Find the forces.

(9) The velocity of a particle in a central orbit varies as $\dfrac{1}{r^2}$. Apply the principle of Least Action to find the orbit, and thence the law of force.

(10) Shew that the amount of heat and light received by a planet in one revolution is inversely as the square root of the latus rectum of its orbit.

(11) If P, P' be the central forces for an orbit and its hodograph,

$$PP' = \frac{h'^2}{h^2}\,rr'.$$

(12) The hodograph is a circle about a point in its circumference, and if θ be the angle which the radius vector makes with the diameter, the angular velocity is given by

$$\frac{d\theta}{dt} = \frac{k}{\sqrt{(\epsilon^{2kt} - 1)}};$$

shew that the path is a cycloid with its vertex upwards, and the velocity at any point, that due to a fall from the tangent at the vertex.

(13) The hodograph for a particle moving in a vertical circle with the velocity due to the depth below the highest point, is

$$r = c \cos \frac{\theta}{2}.$$

(14) When the hodograph is a straight line described uniformly, the path is the trajectory of a projectile in vacuo.

(15) When it is a straight line described with uniform angular velocity about a point, the path is the catenary of uniform strength,

$$\epsilon^{ky} = \sec kx.$$

(16) The hodograph for a circle about a point in the circumference, is a parabola about the focus described with angular velocity proportional to the radius vector.

CHAPTER X.

IMPACT.

207. We come next to the consideration of the effects of a class of forces which cannot be treated by the method just employed. These are called impulsive forces and are such as arise in cases of collision; lasting for an indefinitely short time, and yet producing finite changes of momentum.

When two balls of glass or ivory impinge on one another, no doubt there goes on a very complicated operation during the brief interval of contact. First the portions of the surfaces immediately in contact are disfigured and compressed until the molecular forces thus called into action are sufficient to resist farther compression. At this instant it is evident that the points in contact are moving with the same velocity. But, most substances being endowed with a certain degree of elasticity, the balls tend to recover their spherical form, and an additional force is generated; proportional it is found by experiment to that exerted during the compression. The coefficient of proportionality is a quantity known as the modulus of elasticity. It is always less than 1, as no body is known which possesses perfect elasticity.

The method of treating questions involving forces of this nature, will be best explained by taking as an example the case of direct impact of one spherical ball on another; first, when the balls are inelastic. Again, when their coefficient of elasticity is given.

And it is evident that in the case of direct impact of spheres we may consider the latter as mere particles, since everything is symmetrical about the line joining their centers.

208. Suppose then a sphere of mass M moving with a velocity v, impinge on another of mass M' moving in the same direction with velocity v'; and suppose that at the instant when the mutual compression ceases, the spheres are moving with a common velocity V. If P be the common action

between them at any time t during the compression, it must evidently be of the nature of a moving force exerted by each on the other; and we have if τ be the whole time of compression,

$$M\ (v - V) = \int_0^\tau Pdt = R \text{ suppose},$$

$$M'\ (V - v') = \int_0^\tau Pdt = R;$$

$$\text{and } \therefore\ V = \frac{Mv + M'v'}{M + M'} \text{ and } R = \frac{MM'}{M + M'}\,(v - v').$$

From this result we see that the whole momentum after impact is the same as before, and that the common velocity is that of the center of gravity before impact. Had the balls been moving in opposite directions, v' would have been negative, and

$$V = \frac{Mv - M'v'}{M + M'} \text{ and } R = \frac{MM'}{M + M'}\,(v + v'),$$

whence both balls will be reduced to rest if $Mv = M'u'$; or if the momenta are equal and opposite.

209. This is the general solution of the problem in the case of the balls being inelastic, or having no tendency to recover their original form after compression; but suppose them to be imperfectly elastic. In this case as before noticed, after compression has ceased, the tendency to resume the original form creates an additional action between the balls; proportional to the first. Let e be the coefficient of proportionality, v_1, v_2, the velocities of the balls when finally separated. Then as before

$$M\ (V - v_1) = eR,$$

$$M'\ (v_2 - V) = eR;$$

$$\text{and } \therefore\ Mv_1 = M\,\frac{Mv + M'v'}{M + M'} - e\,\frac{MM'}{M + M'}\,(v - v'),$$

$$v_1 = \frac{(M - eM')\,v + M'\,(1 + e)\,v'}{M + M'} = v - \frac{M'}{M + M'}\,(1 + e)\,(v - v'),$$

with a similar expression for v_2.

A rather singular result may easily be deduced from the last formula. Suppose $M = M'$, $e = 1$, that is, let the balls be of equal mass, and perfectly elastic; then in this case

$$v_1 = v', \text{ and similarly } v_2 = v,$$

or the balls whatever be their velocities interchange these, and the motion is the same as if they had passed through one another without exerting any action whatever.

210. The only other case in which we can at present enter, is that of oblique impact when the balls are perfectly smooth, as in other cases rotations are generated and the motion of each sphere requires to be treated as that of a rigid body.

The simplest case is that of a particle impinging with given velocity, and in a given direction, on a smooth plane.

Suppose the plane of the particle's motion to be taken as that of reference; its trace on the given plane the axis of x, and the point at which the impact takes place, as origin.

The impulsive effect of the plane will evidently be perpendicular to it, since it is smooth. Let this be called R; and let the velocity of the particle be resolved into two v_x, v_y, respectively parallel to the axes. For the first part of the impact,

$$M(v_x - v_x') = 0,$$
$$M(v_y - v_y') = R.$$

But v_y' being the common velocity of the plane and ball is evidently zero;

$$\therefore \; v_x' = v_x, \quad v_y' = 0,$$

or, the velocity parallel to the plane is unchanged, while that perpendicular to it is destroyed. So far for an inelastic ball. If the ball be elastic; let v_x'', v_y'' be the final velocities, then

$$M(v_x' - v_x'') = 0,$$
$$M(v_y' - v_y'') = eR;$$
$$\therefore \; v_x'' = v_x' = v_x,$$

or the velocity parallel to the plane is unaffected; but

$$Mv_y'' = -eR = -eMv_y,$$
$$\text{or, } v_y'' = -ev_y,$$

Hence the velocity perpendicular to the plane is reversed in direction, and diminished in the ratio $e : 1$.

If we designate by the name of angle of incidence the inclination of the original direction of the ball's motion to the normal to the plane, and give that of angle of reflexion to the angle made with the same line by the path after impact; then denoting the total velocities before and after impact by V and V'', and these angles by θ, ϕ respectively, we have

$$V\sin\theta = v_x, \quad V''\sin\phi = v_x'',$$
$$V\cos\theta = v_y, \quad V''\cos\phi = v_y'';$$

∴ by the previous results

$$\left.\begin{aligned} \cot\theta &= -e\cot\phi \\ V'' &= V\frac{\sin\theta}{\sin\phi} \end{aligned}\right\},$$

formulæ sometimes of use.

Of course these results apply to cases of impact on a smooth surface, by making the legitimate hypothesis that the impact and its consequences on the motion of the ball would be the same if for the surface its tangent plane at the point of contact were substituted.

211. PROP. *Two smooth spheres moving in given directions and with given velocities impinge; to determine the impact and the subsequent motion.*

Let the masses of the spheres be M, M'; their velocities before impact v and v', and let the original directions of motion make with the line which joins the centers at the instant of impact, angles α, α'. These angles may easily be calculated from the data, if the radii of the spheres be given.

It is evident that, since the spheres are smooth, the entire impulsive action takes place in the line joining the centers at that instant, and that therefore the future motion of each sphere will be in the plane passing through this line and its original direction of motion.

Let R be the impulse, e the elasticity; then since the velocities in the line of impact are $v\cos\alpha$ and $v'\cos\alpha'$, we have for their final values v_1, v_2, after restitution, by Art. 209, the formulæ

$$v_1 = v\cos\alpha - \frac{M'}{M+M'}(1+e)(v\cos\alpha - v'\cos\alpha'),$$

$$v_2 = v'\cos\alpha' + \frac{M}{M+M'}(1+e)(v\cos\alpha - v'\cos\alpha'),$$

and the value of R is

$$\frac{MM'}{M+M'}(1+e)(v\cos\alpha - v'\cos\alpha').$$

Hence, the sphere M has finally a velocity v_1 in the line joining the centers, and a velocity $v\sin\alpha$ in a known direction perpendicular to this, namely in the plane through this and its original direction of motion. And similarly for the sphere M'. Thus the impact is completely determined.

212. Recurring to the equations in Art. 208, we have

$$M(v - V) = R,$$

$$M'(V - v') = R,$$

and eliminating V, $R = \frac{MM'}{M+M'}(v - v')$......... (1);

$\therefore$ if e be the elasticity, v_1, v_1' the final velocities,

$$\left.\begin{aligned} v_1 &= v - \frac{R(1+e)}{M} \\ v_1' &= v' + \frac{R(1+e)}{M'} \end{aligned}\right\} \text{.................. (2).}$$

Hence, $Mv_1 + M'v_1' = Mv + M'v'$, whatever be e, or there is no momentum lost.

Again, $\quad Mv_1^2 + M'v_1'^2 = Mv^2 + M'v'^2$

$$-2R(1+e)(v-v') + R^2(1+e)^2\frac{M+M'}{MM'}$$

$$= Mv^2 + M'v'^2 - R^2(1+e)^2\frac{M+M'}{MM'}\left(\frac{2}{1+e} - 1\right).$$

The last term of the right-hand side is therefore the vis viva lost by the impact. When $e=0$, its value is a maximum, or $\frac{MM'}{M+M'}(v-v')^2$. When $e=1$, its value is zero, that is, in the direct impact of perfectly elastic bodies, no vis viva is lost.

Also by (2),

$$v_1'-v_1=v'-v+R(1+e)\frac{M+M'}{MM'}$$

$$=e(v-v'), \text{ by } (1).$$

Hence the velocity of separation is e times that of impact.

213. We proceed to some particular problems illustrating this branch of the subject.

PROP. *To one extremity of an uniform and perfectly flexible chain lying in a given curve on a smooth horizontal plane, a given impulsive tension is applied in the direction of the tangent at that extremity; it is required to find the impulsive tension at any other point of the chain.*

Let this be T at a point of the chain whose co-ordinates are x, y; and let the initial velocities of that point parallel to the axes be v_x, v_y, then μ being the mass of a unit of length of the chain, we have the following equations:

$$\left.\begin{aligned}\frac{d}{ds}\left(T\frac{dx}{ds}\right)=\mu v_x\\ \frac{d}{ds}\left(T\frac{dy}{ds}\right)=\mu v_y\end{aligned}\right\}\ldots\ldots\ldots\ldots\ldots\ldots (1).$$

The geometrical condition will be determined as follows. The chain being inextensible, the length of an element δs is invariable, therefore the velocities of its two extremities resolved along it must be the same. This gives evidently

$$\left(\frac{dv_x}{ds}\frac{dx}{ds}+\frac{dv_y}{ds}\frac{dy}{ds}\right)\delta s=0.\ldots\ldots\ldots\ldots\ldots (2).$$

From these equations we proceed to eliminate v_x, v_y.

16—2

By (1) we have by differentiation with respect to s,

$$\mu\frac{dv_x}{ds} = T\frac{d^3x}{ds^3} + 2\frac{dT}{ds}\frac{d^2x}{ds^2} + \frac{d^2T}{ds^2}\frac{dx}{ds},$$

$$\mu\frac{dv_y}{ds} = T\frac{d^3y}{ds^3} + 2\frac{dT}{ds}\frac{d^2y}{ds^2} + \frac{d^2T}{ds^2}\frac{dy}{ds};$$

$\therefore$ multiplying by $\frac{dx}{ds}$, $\frac{dy}{ds}$, respectively, and adding, we have by (2),

$$T\left(\frac{dx}{ds}\frac{d^3x}{ds^3} + \frac{dy}{ds}\frac{d^3y}{ds^3}\right) + 2\frac{dT}{ds}\left(\frac{dx}{ds}\frac{d^2x}{ds^2} + \frac{dy}{ds}\frac{d^2y}{ds^2}\right)$$

$$+ \frac{d^2T}{ds^2}\left\{\left(\frac{dx}{ds}\right)^2 + \left(\frac{dy}{ds}\right)^2\right\} = 0. \quad \ldots\ldots\ldots\ldots (3).$$

But $\left(\frac{dx}{ds}\right)^2 + \left(\frac{dy}{ds}\right)^2 = 1$; differentiate,

$$\therefore \frac{dx}{ds}\frac{d^2x}{ds^2} + \frac{dy}{ds}\frac{d^2y}{ds^2} = 0,$$

and differentiating again, and transposing, we have

$$\frac{dx}{ds}\frac{d^3x}{ds^3} + \frac{dy}{ds}\frac{d^3y}{ds^3} = -\left\{\left(\frac{d^2x}{ds^2}\right)^2 + \left(\frac{d^2y}{ds^2}\right)^2\right\}$$

$$= -\frac{1}{\rho^2},$$

where ρ is the radius of curvature of the element δs.

By means of these transformations (3) takes the final form

$$\frac{d^2T}{ds^2} - \frac{T}{\rho^2} = 0. \quad \ldots\ldots\ldots\ldots\ldots\ldots\ldots (4).$$

This of course cannot be integrated unless the form of the chain is given, or ρ given in terms of s.

To find the instantaneous direction of motion of any point we must find v_x and v_y, and their ratio is the tangent of the angle which the required direction makes with the axis of x. These quantities must be found from the value of T in (4) by means of (1).

214. *Example* I. As a particular example, suppose the chain to form a semicircle of radius a. Then $\rho = a$, and (4) becomes

$$\frac{d^2T}{ds^2} - \frac{T}{a^2} = 0,$$

whose integral is known to be

$$T = A\epsilon^{\frac{s}{a}} + B\epsilon^{-\frac{s}{a}}.$$

To determine the arbitrary constants, observe that when

$$s = 0, \quad T = T_0,$$

the original impulse, and when $s = \pi a$, or at the free extremity of the chain, $T = 0$. Thus we have

$$T_0 = A + B,$$

$$0 = A\epsilon^{\pi} + B\epsilon^{-\pi};$$

$$\therefore \; A = -\frac{T_0\epsilon^{-\pi}}{\epsilon^{\pi} - \epsilon^{-\pi}}, \quad B = \frac{T_0\epsilon^{\pi}}{\epsilon^{\pi} - \epsilon^{-\pi}};$$

$$\therefore \; T = \frac{T_0\left\{\epsilon^{\pi - \frac{s}{a}} - \epsilon^{-\left(\pi - \frac{s}{a}\right)}\right\}}{\epsilon^{\pi} - \epsilon^{-\pi}}$$

$$= T_0\,\frac{\epsilon^{(\pi-\theta)} - \epsilon^{-(\pi-\theta)}}{\epsilon^{\pi} - \epsilon^{-\pi}} \ldots\ldots\ldots\ldots\ldots \;(5),$$

if $s = a\theta$, or we consider the tension at a point whose distance from the tended end subtends an angle θ at the center.

Suppose now that the axis of y is the tangent at the tended end; that of x being the diameter through that point,

$$\text{then } x = a(1 - \cos\theta),$$

$$y = a\sin\theta;$$

$$\therefore \frac{dx}{ds} = \sin\theta,$$

$$\frac{dy}{ds} = \cos\theta;$$

$$\therefore \text{ by (1) and (5) } \frac{v_y}{v_x} = \frac{\dfrac{d}{d\theta}\left[\cos\theta\left\{\epsilon^{\pi-\theta} - \epsilon^{-(\pi-\theta)}\right\}\right]}{\dfrac{d}{d\theta}\left[\sin\theta\left\{\epsilon^{\pi-\theta} - \epsilon^{-(\pi-\theta)}\right\}\right]}.$$

Differentiating out, and then substituting different values of θ, we get the initial directions of corresponding points of the chain. Thus, for the tended end it will easily be seen that putting $\theta = 0$, we have

$$\frac{v_y}{v_x} = -\frac{\epsilon^{\pi} + \epsilon^{-\pi}}{\epsilon^{\pi} - \epsilon^{-\pi}}.$$

For the free end

$$\frac{v_y}{v_x} = \infty,$$

as we should expect, since there is initially no force on it parallel to the axis of x.

215. *Example* II. Suppose it be required that the tension at each point should be proportional to the distance from the free end of the chain.

Then l being the length, and s as before,

$$T = T_0\left(1 - \frac{s}{l}\right) \text{ by hypothesis;}$$

$$\therefore \frac{d^2T}{ds^2} = 0, \quad \text{or by (4)} \ \frac{T}{\rho^2} = 0, \quad \text{and} \ \therefore \rho = \infty,$$

or the chain must lie in a straight line, as is otherwise evident.

216. PROP. *To find the angle which the initial direction of any point makes with the tangent at that point.*

Generally, let $$\frac{v_y}{v_x} = \tan\phi = \frac{\frac{d}{ds}\left(T\frac{dy}{ds}\right)}{\frac{d}{ds}\left(T\frac{dx}{ds}\right)},$$

and $$\frac{\frac{dy}{ds}}{\frac{dx}{ds}} = \tan\psi.$$

Then $(\phi - \psi)$ is the required angle;

$$\therefore \tan(\phi - \psi) = \frac{\frac{d}{ds}\left(T\frac{dy}{ds}\right)\frac{dx}{ds} - \frac{d}{ds}\left(T\frac{dx}{ds}\right)\frac{dy}{ds}}{\frac{d}{ds}\left(T\frac{dx}{ds}\right)\frac{dx}{ds} + \frac{d}{ds}\left(T\frac{dy}{ds}\right)\frac{dy}{ds}}$$

$$= \frac{T\left(\frac{dx}{ds}\frac{d^2y}{ds^2} - \frac{dy}{ds}\frac{d^2x}{ds^2}\right)}{\left(\frac{dT}{ds}\right)}$$

$$= \frac{\left(\frac{T}{\rho}\right)}{\left(\frac{dT}{ds}\right)} \quad \ldots\ldots\ldots\ldots\ldots\ldots (6).$$

Hence again, if every point of the chain is to move initially along the chain $\phi - \psi = 0$, and $\therefore \rho = \infty$, or the chain can only be in a straight line.

217. PROP. *To find the absolute initial velocity of any point of the chain.*

Squaring and adding equations (1) after performing the differentiations indicated, we have

$$\mu^2 (v_x^2 + v_y^2) = \left(\frac{dT}{ds}\right)^2 \left\{\left(\frac{dx}{ds}\right)^2 + \left(\frac{dy}{ds}\right)^2\right\}$$

$$+ 2T \frac{dT}{ds}\left(\frac{dx}{ds}\frac{d^2x}{ds^2} + \frac{dy}{ds}\frac{d^2y}{ds^2}\right)$$

$$+ T^2 \left\{\left(\frac{d^2x}{ds^2}\right)^2 + \left(\frac{d^2y}{ds^2}\right)^2\right\}$$

$$= \left(\frac{dT}{ds}\right)^2 + \frac{T^2}{\rho^2}, \text{ (Art. 213).}$$

Again, if v_s be the velocity resolved along the curve

$$v_s^2 = (v_x^2 + v_y^2) \cos^2 (\phi - \psi)$$

$$= \frac{1}{\mu^2}\left\{\left(\frac{dT}{ds}\right)^2 + \frac{T^2}{\rho^2}\right\} \frac{\left(\frac{dT}{ds}\right)^2}{\left(\frac{dT}{ds}\right)^2 + \frac{T^2}{\rho^2}}$$

$$= \frac{1}{\mu^2}\left(\frac{dT}{ds}\right)^2,$$

which might have been found at once from

$$\mu v_x = \frac{d}{ds}\left(T \frac{dx}{ds}\right),$$

by making the axis of x parallel to the element δs.

The whole problem might have been solved by resolving at any point of the chain, along and perpendicular to the tangent. Calling v_s, v_ρ, the initial velocities in these directions, we have at once

$$\left.\begin{aligned} \delta T &= \mu v_s \delta s \\ \frac{T}{\rho}\delta s &= \mu v_\rho \delta s \end{aligned}\right\},$$

from which the above results may easily be derived.

218. The only other case we shall consider is that of a continuous series of indefinitely small impacts, whose effect is comparable with that of a finite force. The method of considering such a problem is to estimate separately the changes in the velocity produced by the finite forces, and the impacts, in the same indefinitely small time δt, and compound these for the actual effect on the motion in that period.

219. Prop. *A spherical rain-drop descending by the action of gravity, receives continually by precipitation of vapor an accession of mass proportional to its surface; a being its radius when it begins to descend, and r its radius after the interval t, shew that its velocity is given by the equation*

$$v = \frac{gt}{4}\left(1 + \frac{a}{r} + \frac{a^2}{r^2} + \frac{a^3}{r^3}\right),$$

the resistance of the air being left out of account.

Let e be the thickness of the shell of fluid deposited in an unit of time. Then evidently

$$r = a + et \;\ldots\ldots\ldots\ldots\ldots\ldots\ldots\ldots \;(1).$$

Also let $\delta v = \delta_1 v + \delta_2 v$ be the increase of velocity in time δt; the first term due to gravity, the second to the impacts.

Evidently, $\delta_1 v = g\delta t$; and if M be the mass at time t, $\delta(Mv) = 0$ is the condition of the impact.

This gives

$$M\delta_2 v = -v\delta M,$$

$$\text{or } \delta_2 v = -v\,\frac{4\pi r^2 e\delta t}{\frac{4}{3}\pi r^3} = -\frac{3ev\delta t}{r} = -\frac{3ev\delta t}{a+et};$$

$$\therefore \frac{dv}{dt} = g - \frac{3ev}{a+et}.$$

Multiplying by $(a+et)^3$, and transferring the last term to the left hand side of the equation, it gives by inspection

$$(a+et)^3\,v = \frac{g}{4e}(a+et)^4 + C.$$

But $$0 = \frac{g}{4e} a^4 + C \text{ by condition.}$$

$$\therefore\ v = \frac{g}{4e}\left\{(a+et) - \frac{a^4}{(a+et)^3}\right\}.$$

Substituting for *e* from (1),

$$v = \frac{gt}{4(r-a)}\left(r - \frac{a^4}{r^3}\right)$$

$$= \frac{gt}{4}\left(1 + \frac{a}{r} + \frac{a^2}{r^2} + \frac{a^3}{r^3}\right),$$

as required.

220. PROP. *One end B of an uniform heavy chain hangs over a small smooth pully A, and the other is coiled up on a table at C. If B preponderates, determine the motion.*

The moving force due to gravity is the weight of AB minus that of $AC = \mu g(x-a)$ suppose.

Now in an indefinitely small interval δt, this would generate in the portion BAC of the chain an increment of velocity $\delta_1 v$

$$= \frac{\mu g(x-a)}{\mu(x+a)}\delta t. \quad \text{(See Chap. XII.)}$$

But the whole uncoiled chain being in motion at the commencement of the interval δt with velocity v lifts up a portion of length $v\delta t$ from the table during that interval. Hence if $\delta_2 v$ be the change of velocity arising from this impact, we have by the condition that no momentum is lost, i. e.,

$$V' = \frac{MV}{M+M'},$$

$$v + \delta_2 v = \frac{\mu(x+a)v}{\mu(x+a) + \mu v\delta t},$$

$$\text{or } \delta_2 v = -\frac{v^2\delta t}{x+a},$$

omitting quantities of the second and higher orders.

Hence as $\frac{\delta v}{\delta t} = \frac{\delta_1 v}{\delta t} + \frac{\delta_2 v}{\delta t}$,

proceeding to the limit

$$\frac{dv}{dt} = v\frac{dv}{dx} = \frac{g(x-a) - v^2}{(x+a)};$$

$$\therefore\ (x+a)^2 v \frac{dv}{dx} + v^2 (x+a) = g(x^2 - a^2)$$

$$\text{or } (x+a)^2 v^2 = (x+a)^2 \left(\frac{dx}{dt}\right)^2 = \int 2g(x^2 - a^2)\, dx,$$

which from given initial circumstances determines the velocity at any instant.

EXAMPLES.

(1) One perfectly elastic ball cannot be reduced to rest by direct impact on another equal ball, unless the latter is at rest.

(2) If two perfectly elastic balls impinge directly with equal velocities, their masses must be as 1 : 3 that one may be reduced to rest.

(3) Shew that if two equal balls of elasticity e impinge directly with velocities $\frac{1+e}{1-e}V$ and $-V$, the latter will be reduced to rest.

(4) Shew that the mass of the ball which must be interposed directly between M at rest, and M' moving with a given velocity V, so that M may acquire the greatest velocity, is

$$\sqrt{(MM')},$$

and that that maximum velocity is $\frac{M'V(1+e)^2}{\{\sqrt{(M)} + \sqrt{(M')}\}^2}$.

(5) Suppose the balls perfectly elastic, and an infinite number to be interposed, shew that the maximum velocity which can thus be given to M, is

$$V\sqrt{\left(\frac{M'}{M}\right)}.$$

Note that by the preceding question, the masses must form a geometric series, and the above result is easily deduced.

(6) Perfectly elastic particles slide down radii vectores from the focus of a parabola whose axis is horizontal and plane vertical. After reflexion at the curve they describe their trajectories. What is the locus of the foci?

(7) A impinges on B, elasticity being e, shew that A's deviation is greatest when its tangent is $\frac{1}{2}\frac{1+e}{1-e}\sqrt{\left(\frac{1-e}{2}\right)}$.

(8) A particle of elasticity $\frac{1}{m}$ is projected from a point in a smooth horizontal plane. Find how far it goes before it ceases to rebound. Shew that the times between successive rebounds are in a geometric series whose ratio is $\frac{1}{m}$, and the heights above the plane in another whose ratio is $\frac{1}{m^2}$.

(9) A perfectly elastic particle is projected from the foot of an inclined plane in a direction making an angle β with the plane; the plane is inclined at an angle α to the horizon. Shew that if $2\tan\beta = \tan\alpha$, the particle will return after one rebound to the point of projection.

If there be two rebounds before coming back to the point of projection, and the elasticity be e,

$$\cot\beta = (1 + e + e^2)\tan\alpha.$$

(10) AOB is the vertical diameter of a circle. A perfectly elastic particle slides down any chord AC, and is reflected at BC. The locus of the focus of its path is the circle whose diameter is AO.

(11) If the direction in which one ball is moving when it impinges on another equal ball at rest, bisect the angle between their future directions; then that angle is

$$2 \tan^{-1} \sqrt{(e)}.$$

(12) If $e = \frac{1}{2}$, find the direction in which a ball must be projected against a smooth vertical wall, so as with the least possible velocity, to return to the point of projection.

(13) ABC is a triangle, a, b, c the points of contact of the inscribed circle with the sides, a being in BC, &c. Shew that if a particle of elasticity e projected from a to b be reflected to c, $Ab = e\,Cb$, and if it return to Da,

$$AB = e\,AC.$$

(14) A number of balls of given elasticity, A, B, C, &c. are placed in a line, A is projected with given velocity so as to impinge on B, B then impinges on C, and so on; find the masses of the balls B, C, &c. in order that each of the balls A, B, C, &c. may be reduced to rest by impinging on the next; and find the velocity of the nth ball after its impact with the $(n-1)^{\text{th}}$.

(15) An imperfectly elastic ball is projected in a given direction within a fixed horizontal hoop, so as to go on rebounding from the surface of the hoop; find the limit to which the velocity will approach, and shew that it attains this limit in a finite time.

(16) A given inelastic mass is let fall from a given height on one scale of a balance, and two inelastic masses are let fall from different heights on the other scale, so that the three impacts take place simultaneously; find the relations between the masses and heights that the balance may remain permanently at rest.

(17) A particle moving in a parabola about a center of force in the focus, strikes a hard plane at any point of its path. If it describe a parabola after the impact, find the direction of its axis.

(18) OA, OB are rods in the same vertical plane inclined at angles α, β to the horizon. If a perfectly elastic particle projected from A, strike B and continue to rebound between A and B, then T being the time of flight, γ the inclination of AB to the horizon,

$$\tan\gamma = \tfrac{1}{2}(\cot\beta - \cot\alpha),$$

$$\text{and } T^2 = \frac{2c}{g}\,\frac{\sin(\alpha+\beta)}{\sqrt{\{4\sin^2\alpha\sin^2\beta + \sin^2(\alpha-\beta)\}}}.$$

(19) Equal particles of elasticity e revolve in opposite directions about the focus in an ellipse of excentricity $\frac{3}{5}$, and impinge at the nearer apse. Find the distances of future impacts, and shew that if p be the original apsidal distance, the particles fall into the center after the time

$$\frac{\pi}{14}\,\frac{(5p)^{\frac{3}{2}}}{\sqrt{(2\mu)}}.$$

(20) Two equal particles, connected by a string which passes freely through the pole, are constrained to move in the same logarithmic spiral. If they be originally at rest and one be projected with given velocity (so as to increase its distance from the pole), determine the impact.

(21) Three perfectly elastic balls (supposed indefinitely small) are placed at the corners of a triangle. To find the relations among the masses that the sphere A if projected to strike B, may be reflected to C and from C to its original position. The impacts are supposed to take place at each corner so that the line joining the centers of the spheres is perpendicular to the opposite side of the triangle; α, β, γ, being the angles of the triangle, we find

$$\frac{A}{B} = \frac{\sin\beta}{\sin(\alpha-\gamma)}, \quad \frac{A}{C} = \frac{\sin\gamma}{\sin(\beta-\alpha)}.$$

(22) If a rocket of mass M throw off every unit of time a mass eM with relative velocity V, and if M' be the weight of the case, &c., shew that it cannot rise at once unless $Ve > g$, nor at all unless $\frac{MVe}{M'} > g$. If it do rise at once vertically, shew that its greatest velocity is

$$V \log \frac{M}{M'} - \frac{g}{e}\left(1 - \frac{M'}{M}\right),$$

and the greatest height it reaches

$$\frac{V^2}{2g}\left(\log \frac{M}{M'}\right)^2 + \frac{V}{e}\left(1 - \frac{M'}{M} - \log \frac{M}{M'}\right).$$

(23) If an infinite number of perfectly elastic material points equally distributed through a hollow sphere, be set in motion each with any velocity, shew that the resulting continuous pressure (referred to a unit of area) on the internal surface is equal to one third of the *vis viva* of the particles divided by the volume of the sphere.

(24) A comet in moving from one given point to another, throws off at every instant small portions of its mass which always bear the same ratio n to the mass which remains. If v be the velocity with which each particle is thrown off, α the inclination of its direction to the radius vector, prove that the period t will be diminished by

$$\frac{3nvt}{fa}\{(\phi' - \phi)\sqrt{(ap)}\sin\alpha - (r' - r)\cos\alpha\},$$

ϕ and ϕ' being the excentric anomalies, r and r' the focal distances at the given points, a the mean distance, $2p$ the latus rectum, and f the force at distance a.

CHAPTER XI.

DISTURBED MOTION.

221. In the investigation of the motion of a particle subjected to the action of disturbing forces, we may, when the latter are very small in comparison with the forces under which the undisturbed orbit is described, suppose that at any instant the actual orbit is of the same *nature* as the undisturbed, but that its magnitude, form and position are slightly different. By this means the consideration of motion in an orbit whose equation cannot be found, or if found would be of extreme complexity, is reduced to the cases considered in the foregoing Chapters, the only additional process being the determination of the changes of the elements or *Parameters* of the orbit, due to the disturbing forces; thus making the parameters explicit functions of the time. The principal use of this method is in the planetary theory, and there the elements of the instantaneous orbit cannot be determined but by approximation, which this method affords us the best means of effecting.

It is not necessary that the *orbit* should be changed by the disturbing force, in order that the method of parameters be applicable; suppose for instance a particle be constrained to move on a given curve; the extent of its oscillations, or the velocity with which it reaches a particular point in the path, are parameters, and in terms of such the motion may be expressed. This method is therefore applicable to any case of free or constrained motion, always supposing the disturbing forces to be small; the only difference being that in constrained motion there are fewer parameters of which to find the variation.

222. The general principle of the method may be explained as follows.

Let
$$\left.\begin{aligned}\frac{d^2x}{dt^2} &= X + X' \\ \frac{d^2y}{dt^2} &= Y + Y' \\ \frac{d^2z}{dt^2} &= Z + Z'\end{aligned}\right\} \ldots\ldots\ldots\ldots\ldots\ldots (1),$$

be the equations of motion of a particle whose co-ordinates at time t are xyz; and suppose farther, that X', Y', Z' are the sums of the resolved parts of the disturbing forces parallel to the axes. Had there been no disturbing forces, we should have had the equations of motion

$$\left.\begin{aligned}\frac{d^2x}{dt^2} &= X \\ \frac{d^2y}{dt^2} &= Y \\ \frac{d^2z}{dt^2} &= Z\end{aligned}\right\} \ldots\ldots\ldots\ldots\ldots\ldots\ldots\ldots (2).$$

Now let the solution of equations (2) be

$$\left.\begin{aligned}x &= \phi\,(a,\ \alpha,\ t) \\ y &= \chi\,(b,\ \beta,\ t) \\ z &= \psi\,(c,\ \gamma,\ t)\end{aligned}\right\} . \ldots\ldots\ldots\ldots\ldots\ldots (3),$$

each involving two arbitrary constants.

Let us remark in passing, that, if the motion were constrained, we should have in addition one or two relations between x, y and z, so that the *number* of arbitrary constants would be reduced.

Now suppose the solutions of (1) to be of the same form as the expressions in (3), a, α, b, β, c, γ being no longer constants but functions of t to be determined.

The fact of (3) having to satisfy (1), gives us the three following equations for the determination of these parameters,

$$\left.\begin{aligned}\frac{d^2\phi}{dt^2} &= X + X' \\ \frac{d^2\chi}{dt^2} &= Y + Y' \\ \frac{d^2\psi}{dt^2} &= Z + Z'\end{aligned}\right\} \quad\ldots\ldots\ldots\ldots\ldots\ldots (4).$$

223. We are at liberty then to make any other three hypotheses regarding them that we please. The most convenient are those afforded by the condition that not only the expressions for the co-ordinates of the particle, but in addition, the expressions for the resolved parts of the velocity parallel to the co-ordinate axes, should be of the same *form* in the disturbed as in the undisturbed orbit.

Now in the disturbed orbit,

$$\frac{dx}{dt} = \left(\frac{d\phi}{dt}\right) + \left(\frac{d\phi}{d\alpha}\right)\frac{d\alpha}{dt} + \left(\frac{d\phi}{da}\right)\frac{da}{dt},$$

with similar expressions for $\frac{dy}{dt}$ and $\frac{dz}{dt}$, the brackets being used to express *partial* differentiation.

But in the undisturbed orbit

$$\frac{dx}{dt} = \left(\frac{d\phi}{dt}\right);$$

and similarly for $\frac{dy}{dt}$ and $\frac{dz}{dt}$.

From this we have the three additional relations between

a, α, b, β, c, γ and t,

$$\left.\begin{aligned}\left(\frac{d\phi}{d\alpha}\right)\frac{d\alpha}{dt} + \left(\frac{d\phi}{da}\right)\frac{da}{dt} &= 0 \\ \left(\frac{d\chi}{d\beta}\right)\frac{d\beta}{dt} + \left(\frac{d\chi}{db}\right)\frac{db}{dt} &= 0 \\ \left(\frac{d\psi}{d\gamma}\right)\frac{d\gamma}{dt} + \left(\frac{d\psi}{dc}\right)\frac{dc}{dt} &= 0\end{aligned}\right\} \quad\ldots\ldots\ldots\ldots (5),$$

And taking these into account as well as equations (2), (3), equations (4) become

$$\left.\begin{aligned}\left\{\frac{d\left(\frac{d\phi}{dt}\right)}{d\alpha}\right\}\frac{d\alpha}{dt}+\left\{\frac{d\left(\frac{d\phi}{dt}\right)}{da}\right\}\frac{da}{dt}&=X'\\ \left\{\frac{d\left(\frac{d\chi}{dt}\right)}{d\beta}\right\}\frac{d\beta}{dt}+\left\{\frac{d\left(\frac{d\chi}{dt}\right)}{db}\right\}\frac{db}{dt}&=Y'\\ \left\{\frac{d\left(\frac{d\psi}{dt}\right)}{d\gamma}\right\}\frac{d\gamma}{dt}+\left\{\frac{d\left(\frac{d\psi}{dt}\right)}{dc}\right\}\frac{dc}{dt}&=Z'\end{aligned}\right\}\quad\ldots\ldots\ (6).$$

Equations (5) and (6) suffice to determine the six parameters in terms of t, and it may be remarked that, should any of these quantities themselves appear in the differential equations they may be treated as constants, since their variations on account of the smallness of the disturbing forces may be neglected for a short time at any period of the motion. This will evidently amount analytically to neglecting higher powers of the disturbing forces than the first.

224. Supposing then the parameters found as functions of t, if in equations (3) we substitute these values; the result of the subsequent elimination of t will evidently be the two equations to the orbit *actually* described by the particle.

But if t be eliminated considering a, α, &c. constant, and *then* their values as functions of t be substituted; the two resulting equations will for any particular value t' given to t, represent a curve which evidently coincides with the actual path at the time t', and which *would* be from that instant the actual path if the disturbing forces were then to cease. This is called the *instantaneous orbit* at the time $t=t'$, and its form and position are evidently slowly changing.

225. If the constants be fewer than six as in the case of constrained motion, so many equations as those just indicated will not be necessary; but the preceding sketch will enable us to apply the method to any such case. Thus if the particle

be constrained to move on a given surface, taking its equation along with equations (3), there will evidently be two of the six constants at once determined completely in terms of the others. And if the motion be on a given curve, four of the constants will be got rid of. In that case taking tangential resolutions we have, s being the length of the arc described at time t, S the tangential accelerating effect, S' the tangential disturbing force,

$$\frac{d^2s}{dt^2} = S + S' \dots\dots\dots\dots\dots\dots \quad (7).$$

Now if $s = \phi(a, \alpha, t) \dots (8)$ be the solution of

$$\frac{d^2s}{dt^2} = S,$$

we shall get the values of a, α in terms of t, in order that (8) may satisfy (7), and also that the expression for the velocity in the orbit be the same in both cases, by solving the equations,

$$\left(\frac{d\phi}{da}\right)\frac{da}{dt} + \left(\frac{d\phi}{d\alpha}\right)\frac{d\alpha}{dt} = 0,$$

$$\left\{\frac{d\left(\frac{d\phi}{dt}\right)}{da}\right\}\frac{da}{dt} + \left\{\frac{d\left(\frac{d\phi}{dt}\right)}{d\alpha}\right\}\frac{d\alpha}{dt} = S',$$

which evidently correspond to those of groups (5) and (6) respectively.

226. Prop. *To determine the effect of a small disturbing force on a simple cycloidal pendulum.*

If s be the arcual distance of the particle from the lowest point at time t we have (Art. 131), putting

$$\frac{g}{4a} = n^2,$$

$$\frac{d^2s}{dt^2} + n^2 s = 0,$$

as the equation of undisturbed motion. Its solution is

$$s = a \sin (nt + b),$$

where a and b are arbitrary constant quantities depending on the length of the arc of vibration and the time of passing the lowest point.

The velocity at time $t = \frac{ds}{dt} = na \cos (nt + b)$.

We shall now suppose that f is a small disturbing accelerating force resolved along the tangent: the equation of motion then is

$$\frac{d^2s}{dt^2} + n^2 s = f.$$

The solution of this equation we assume to be

$$s = a \sin (nt + b),$$

a and b being considered unknown functions of t, which it is our business now to determine.

Taking as a condition, that the form of the expression for the velocity is to be still the same; since we have

$$\frac{ds}{dt} = na \cos (nt + b) + \frac{da}{dt} \sin (nt + b) + a \cos (nt + b) \frac{db}{dt};$$

$$\therefore \frac{da}{dt} \sin (nt + b) + a \cos (nt + b) \frac{db}{dt} = 0,$$

this is the assumed relation between a and b.

$$\text{Again since } \frac{ds}{dt} = na \cos (nt + b);$$

$$\therefore \frac{d^2s}{dt^2} = -n^2 a \sin (nt + b) + n \frac{da}{dt} \cos (nt + b) - na \sin (nt + b) \frac{db}{dt},$$

in this substitute for $\frac{d^2s}{dt^2}$ its value;

$$\therefore\ n\frac{da}{dt}\cos(nt+b) - na\sin(nt+b)\frac{db}{dt} = f,$$

this is the second equation between a and b.

Eliminating successively $\frac{db}{dt}$ and $\frac{da}{dt}$ from these, we have

$$\frac{da}{dt} = \frac{f}{n}\cos(nt+b), \quad \frac{db}{dt} = -\frac{f}{na}\sin(nt+b).$$

If we could solve these equations we should have the complete determination of the motion. In few cases is this practicable: in all to which we shall have to apply the investigation an approximation is sufficient.

We suppose f to be a very small force. Hence the variable parts of a and b are of the same order of magnitude as f, and may be neglected on the right-hand side of the above equations if we agree to neglect the square and higher powers of f.

In order to find the alteration in the extent of vibration which takes place in one oscillation we must integrate $\frac{f}{n}\cos(nt+b)\,dt$ between the limits of t corresponding to one oscillation; that is, from a value of t which gives $nt+b=\alpha$ to the value of t which gives $nt+b=\pi+\alpha$. Here α may be any quantity: in different cases we shall find it convenient to integrate between different limits.

$$\therefore\ \textit{increase of arc of semi-vibration} = \frac{1}{n}\int f\cos(nt+b)\,dt$$

between the above-mentioned limits.

To find the alteration in the time of oscillation, let T, T' be the values of t at two successive arrivals of the pendulum at the lowest point; B, B' the values of b at these times. Then

$$nT + B = m\pi, \quad nT' + B' = (m+1)\pi;$$

$$\therefore\ n(T'-T) + B' - B = \pi,$$

$$T' - T = \frac{\pi}{n} - \frac{1}{n}(B'-B).$$

$$\text{Now } B' - B = \int_T^{T'} \frac{db}{dt}\,dt = -\frac{1}{na}\int_T^{T'} f \sin(nt+b)\,dt:$$

therefore the increase of time of oscillation

$$= \frac{1}{n^2 a}\int_T^{T'} f \sin(nt+b)\,dt,$$

and *the proportionate increase of time of oscillation*

$$= \frac{1}{\pi na}\int_T^{T'} f \sin(nt+b)\,dt.$$

If the circumstances are such that we must integrate through two vibrations, then

$$\text{proportionate increase of time of osc}^{\text{n}}. = \frac{1}{2\pi na}\int f \sin(nt+b)\,dt.$$

227. These formulæ are convenient when f can be expressed in terms of t. If however f be expressed in terms of s, as is the case particularly in clock escapements, we must modify the formulæ:

$$\frac{da}{ds} = \frac{da}{dt}\,\frac{dt}{ds} = \frac{1}{na\cos(nt+b)}\,\frac{da}{dt} = \frac{f}{n^2 a},$$

$$\text{and } \frac{db}{ds} = \frac{1}{na\cos(nt+b)}\,\frac{db}{dt}$$

$$= -\frac{f}{n^2 a^2}\tan(nt+b) = -\frac{f}{n^2 a^2}\,\frac{s}{\sqrt{(a^2-s^2)}};$$

$$\therefore \textit{ increase of arc of semi-vibration} = \frac{1}{n^2 a}\int_{-s}^{s} f ds,$$

$$\textit{proportionate increase of the time of vib}^{\text{n}}. = \frac{1}{\pi n^2 a^2}\int_{-s}^{s} \frac{fs ds}{\sqrt{(a^2-s^2)}}.$$

The limits should strictly be $-s$ and s', where s' differs from s by a quantity, which depends upon the change in the arc of vibration: but we may neglect this difference between s and s', since the terms in which they occur are small.

228. *Instead of vibrating in a cycloid let the pendulum vibrate in a circle.*

Here the force $= g \sin \frac{s}{l} = \frac{gs}{l} - \frac{gs^3}{6l^3}$ nearly;

$$\therefore f = \frac{g}{6l^3} s^3 = \frac{ga^3}{6l^3} \sin^3 (nt + b);$$

therefore proportionate increase in time of vibration

$$= \frac{ga^2}{6\pi n l^3} \int \sin^4 (nt + b)\, dt.$$

Now $\int \sin^4 (nt + b)\, dt = \frac{1}{8} \int \{3 - 4 \cos 2 (nt + b) + \cos 4 (nt + b)\}\, dt$

$$= \frac{1}{8} \{3t - \frac{2}{n} \sin 2 (nt + b) + \frac{1}{4n} \sin 4 (nt + b)\} + C$$

$$= \frac{3}{8} \frac{\pi}{n}, \text{ from } nt + b = 0 \text{ to } \pi;$$

therefore proportionate increase of time

$$= \frac{ga^2}{16n^2 l^3} = \frac{a^2}{16l^2} \text{ since } n^2 = \frac{g}{l}.$$

The increase of arc of vibration

$$= \frac{ga^3}{6nl^3} \int \cos (nt + b) \sin^3 (nt + b)\, dt$$

$$= \frac{ga^3}{24n^2 l^3} \sin^4 (nt + b) + C = 0 \text{ between the limits,}$$

as we might easily have foreseen.

229. *Suppose the friction at the point of suspension to be constant.*

It will be convenient to take the integrals during that time in which the friction acts in the same direction: that is, from the beginning of a vibration to its end, or from $nt+b=-\frac{1}{2}\pi$ to $nt+b=\frac{1}{2}\pi$. Here $f=-c$, since the friction *retards* the motion;

$$\therefore \text{ increase of arc} = -\frac{c}{n}\int \cos(nt+b)\,dt$$

$$= -\frac{c}{n^2}\sin(nt+b) + C = -\frac{2c}{n^2},$$

$$\text{proportionate increase of time} = -\frac{c}{\pi na}\int \sin(nt+b)\,dt$$

$$= \frac{c}{\pi n^2 a}\cos(nt+b) + C = 0, \text{ between the limits.}$$

230. *Suppose the resistance of the air to produce a force varying as the mth power of the velocity or $=kv^m$, m being any whole number.*

The velocity in moving from the lowest point

$$= na\cos(nt+b);\quad \therefore f = -kn^m a^m \cos^m(nt+b);$$

therefore increase of arc

$$= -kn^{m-1}a^m \int \cos^{m+1}(nt+b)\,dt \text{ from } nt+b=-\tfrac{1}{2}\pi \text{ to } \tfrac{1}{2}\pi$$

$$= -k\pi n^{m-2}a^m \frac{m(m-2)\ldots\ldots\ldots 1}{(m+1)(m-1)\ldots 2}\ (m \text{ odd})$$

$$= -2kn^{m-2}a^m \frac{m(m-2)\ldots\ldots\ldots 2}{(m+1)(m-1)\ldots 3}\ (m \text{ even}).$$

When $m=2$ (the law usually taken) the decrease of the arc $=\frac{4}{3}ka^2$.

The proportionate increase of time of oscillation

$$= -\frac{k}{\pi}n^{m-1}a^{m-1}\int \cos^m(nt+b)\sin(nt+b)\,dt$$

$$= \frac{kn^{m-2}a^{m-1}}{\pi(m+1)}\cos^{m+1}(nt+b) + C = 0, \text{ between limits,}$$

whether m be a positive integer or fraction.

231. *Suppose the resistance of the air is expressed by any function of the velocity.*

Here $f = -\phi(v)$ for motion in the positive direction: and the increase of the arc of vibration

$$= \frac{1}{n^3 a} \int \phi(v) \frac{\cos(nt+b)}{\sin(nt+b)} dv = \frac{1}{n^3 a} \int \frac{v\phi(v)\, dv}{\sqrt{(n^2a^2 - v^2)}}$$

from $v=0$ to $v=0$ again. But it must be observed that from $v=0$ to $v=na$ (that is, from $s=-a$ to $s=0$) the radical must be taken with a negative sign, because $\sin(nt+b)$ is then negative. The increase of the arc is consequently

$$= -\frac{1}{n^3 a} \int_0^{na} \frac{v\phi(v)\, dv}{\sqrt{(n^2a^2 - v^2)}} + \frac{1}{n^3 a} \int_{na}^0 \frac{v\phi(v)\, dv}{\sqrt{(n^2a^2 - v^2)}},$$

and therefore decrease $= \dfrac{2}{n^3 a} \displaystyle\int_0^{na} \frac{v\phi(v)\, dv}{\sqrt{(n^2a^2 - v^2)}}$.

The proportionate increase of time of vibration

$$= -\frac{1}{\pi na} \int \phi(v) \sin(nt+b)\, dt = \frac{1}{\pi n^3 a^2} \int \phi(v)\, dv$$

$$= \frac{1}{\pi n^3 a^2} \psi(v) = 0, \text{ from } v=0 \text{ to } v=0.$$

Hence a resistance which is constant, or which depends on the velocity, does not alter the time of vibration.

232. *Let the resistance be that produced by a current of air moving in the plane of vibration with a velocity V greater than the greatest velocity of the pendulum: and varying as the square of their relative velocity.*

Here $f = \phi(v) = k(V-v)^2$ when the pendulum moves in the direction of the current, which we suppose to be the *positive* direction of s; and $f = \phi(v) = k(V+v)^2$ when it moves in the opposite direction.

By the formula in the last Example, when the pendulum moves in the direction of the current, the arc is increased by

$$k\left(\frac{2V^2}{n^2} - \frac{Va\pi}{n} + \frac{4a^2}{3}\right),$$

and when it returns the arc is diminished by

$$k\left(\frac{2V^2}{n^2}+\frac{Va\pi}{n}+\frac{4a^2}{3}\right).$$

The diminution in two vibrations $=\dfrac{2kVa\pi}{n}$. The time is unaffected.

233. *Let a force F act through a very small space x at the distance c from the lowest point.*

The increase of the arc $=\dfrac{1}{n^2a}\displaystyle\int_c^{c+x}Fds=\dfrac{Fx}{n^2a}$ nearly.

The proportionate increase of the time of vibration

$$=\frac{1}{\pi n^2a^2}\int_c^{c+x}\frac{Fsds}{\sqrt{(a^2-s^2)}},$$

if the general value of the integral be $\phi(s)$, then the proportionate increase of time

$$=\phi(c+x)-\phi(c)=\phi'(c)\,x$$

$$=\frac{Fx}{\pi n^2a^2}\frac{c}{\sqrt{(a^2-c^2)}}.$$

If then an impulse be given when the pendulum is at its lowest point, $c=0$ and the time of vibration is unaffected.

234. PROP. *To determine the motion of a projectile in an uniform medium, the resistance being as the square of the velocity.*

Here as before (Art. 176),

$$\left.\begin{aligned}\frac{d^2x}{dt^2}&=-kv^2\frac{dx}{ds}=-k\frac{ds}{dt}\frac{dx}{dt}\\ \frac{d^2y}{dt^2}&=-g'-kv^2\frac{dy}{ds}=-g'-k\frac{ds}{dt}\frac{dy}{dt}\end{aligned}\right\}\ldots\ldots\ldots\ldots(1).$$

Now if $k=0$, we have evidently

$$\left.\begin{aligned}x&=a+mt\\ y&=b+nt-\tfrac{1}{2}g't^2\end{aligned}\right\}\ldots\ldots\ldots\ldots\ldots\ldots(2),$$

where a, b are the co-ordinates of the point, and m, n the horizontal and vertical velocities of projection.

Now supposing (2) to be the solution of (1), a, b, m and n being now functions of t, we have

$$\left.\begin{aligned}\frac{dx}{dt} &= m,\\ \frac{dy}{dt} &= n - g't,\end{aligned}\right\}$$

in both orbits; therefore

$$\left.\begin{aligned}\frac{da}{dt} + t\frac{dm}{dt} &= 0\\ \frac{db}{dt} + t\frac{dn}{dt} &= 0\end{aligned}\right\}\ldots\ldots\ldots\ldots\ldots\ldots\ldots(3);$$

$$\therefore \frac{ds}{dt} = \sqrt{\{m^2 + (n - g't)^2\}},$$

and equations (1) become

$$\frac{d^2x}{dt^2} = \frac{dm}{dt} = -km\sqrt{\{m^2 + (n - g't)^2\}},$$

$$\frac{d^2y}{dt^2} + g' = \frac{dn}{dt} = -k(n - g't)\sqrt{\{m^2 + (n - g't)^2\}}\ \ldots\ldots\ (4);$$

and (3) becomes

$$\frac{da}{dt} = kmt\sqrt{\{m^2 + (n - g't)^2\}},$$

$$\frac{db}{dt} = k(n - g't)\,t\sqrt{\{m^2 + (n - g't)^2\}}\ \ldots\ldots\ldots\ldots\ (5).$$

The last four equations suffice to determine a, b, m, n in terms of t, and thence the instantaneous orbit. For a first approximation, we may by the remark in Art. 223 integrate these equations on the supposition that the right-hand side of each is variable only as far as it explicitly involves t.

Now by Chapter IV. it is evident that the latus rectum of the instantaneous parabola is $\frac{2m^2}{g'}$, and as $\frac{dm}{dt}$ is negative, we see that this quantity is continually diminishing.

Also by equations (2),

$$\left(x-a-\frac{mn}{g'}\right)^2=-\frac{2m^2}{g'}\left(y-b-\frac{n^2}{2g'}\right).$$

These give for the co-ordinates of the focus of the instantaneous orbit

$$\left.\begin{aligned}x'&=a+\frac{mn}{g'}\\ y'&=b+\frac{n^2-m^2}{2g'}\end{aligned}\right\}.$$

If these expressions be differentiated and $\frac{da}{dt}$, &c. be eliminated by means of (4) and (5), it will be seen that $\frac{dx'}{dt}$ is negative, or the axis of the instantaneous orbit moves backwards until the particle reaches the vertex; after which it progresses for the rest of the motion; also that $\frac{dy'}{dt}$ is positive if $m > n$, that is, the focus of the instantaneous orbit moves upwards while the direction of motion of the particle makes with the horizon an angle less than 45°, i. e. while the particle is above the latus rectum of the instantaneous orbit.

235. Prop. *A particle moving about a center of force whose intensity is inversely as the square of the distance, is subjected to a small disturbing force in its plane of motion; to investigate the change in the form and position of the orbit.*

Let the disturbing force be resolved into two ϕ and ϕ', one along the radius vector and the other perpendicular to it; the equations of motion are

$$\left.\begin{aligned}\frac{d^2r}{dt^2}-r\left(\frac{d\theta}{dt}\right)^2&=-\frac{\mu}{r^2}+\phi\\ \frac{1}{r}\frac{d}{dt}\left(r^2\frac{d\theta}{dt}\right)&=\phi'\end{aligned}\right\}\text{..............}(1).$$

Now the solutions of these equations are

$$\left.\begin{aligned}\frac{1}{r} &= \frac{\mu}{h^2}\{1+e\cos(\theta-\varpi)\} \\ r^2\frac{d\theta}{dt} &= h\end{aligned}\right\}\ldots\ldots\ldots\ldots\ldots (2),$$

if we omit the forces ϕ and ϕ'. When we consider their effect then, the quantities h, e, and ϖ must be considered variable.

But in the instantaneous orbit, the velocity and direction of motion are the same as in the actual orbit, and therefore if (2) be differentiated, considering h, e, and ϖ variable, the results for r, $\frac{dr}{d\theta}$, and $\frac{d\theta}{dt}$ must be in the same form as if the disturbing forces had not acted. This will enable us to avoid second differential coefficients of h, e and ϖ; and the substitution for $\frac{dr}{dt}$, $\frac{d^2r}{dt^2}$, and $\frac{d\theta}{dt}$ in (1), will give us altogether three equations for

$$\frac{dh}{dt},\ \frac{de}{dt},\ \frac{d\varpi}{dt}.$$

The expressions for these quantities are complicated and so we do not give them. They will be more easily investigated in particular cases, when ϕ and ϕ' are given.

In the case of the orbit being an ellipse $h^2 = \mu a(1-e^2)$, so that we have by substitution

$$\frac{da}{dt},\ \frac{de}{dt},\ \text{and}\ \frac{d\varpi}{dt}.$$

Also the second integral of the second of equations (1) involves ϵ or the epoch, which will also be thus found as a function of t.

236. If we desire the change produced in the form and position of an orbit by a slight change made in the velocity, or direction of motion, &c. at some particular point, we must express the elements of the orbit in terms of the quantity to be changed; then taking the differentials of both sides, we have the required value.

Thus, we have generally in an elliptic orbit

$$v^2 = \mu\left(\frac{2}{r} - \frac{1}{a}\right).$$

At the extremity of the axis major farthest from the focus this becomes

$$V^2 = \frac{\mu}{a}\,\frac{1-e}{1+e}.$$

Now if at this point V be made $V + \delta V$, we have the condition that in the new orbit $a(1+e)$ shall have the same value as in the old; since this will still be the apsidal distance.

Hence

$$\delta(V^2) = \delta\left(\frac{\mu}{a}\,\frac{1-e}{1+e}\right),$$

$$\text{and } \delta\{a(1+e)\} = 0;$$

$$\therefore\ 2V\delta V = -\frac{\mu}{a}\,\frac{\delta e}{1+e},$$

$$\text{or } \delta e = -2\sqrt{\left\{\frac{a}{\mu}(1-e^2)\right\}}\,\delta V.$$

$$\text{And } \delta a = -\frac{a}{1+e}\,\delta e$$

$$= 2\sqrt{\left(\frac{a^3}{\mu}\,\frac{1-e}{1+e}\right)}\,\delta V,$$

which determine the increase of the axis major and diminution of the excentricity, and the same method is applicable to more complicated cases.

EXAMPLES.

(1) If a small velocity $n\dfrac{\mu e}{h}$ be impressed on a planet, in the direction of the radius vector, shew that

$$\delta e = ne \sin(\theta - \varpi),$$

$$\delta\varpi = -n \cos(\theta - \varpi).$$

(2) A satellite moves about a spherical planet in the plane of its equator, in a slightly elliptic orbit. Find the motion of the apse due to an uniform mountain ridge at the equator.

(3) Central force varying as the distance, the velocity of a particle is increased by $\frac{1}{n}$th when it is at the extremity of one of the equal conjugate diameters of its orbit. Shew that each axis is increased by $\frac{1}{2n}$th, and that the apse regredes through an angle

$$\frac{1}{n}\frac{ab}{a^2-b^2}.$$

(4) At what point of an elliptic orbit described about the focus, can a small change be made in the direction of motion without altering the position of the apse?

If $\delta\phi$ be this change, shew that (in the supposed case)

$$\delta\phi = \frac{\delta e}{1-e^2}.$$

(5) Shew that in an elliptic orbit about the focus, if the velocity be increased by $\frac{1}{n}$th when the true anomaly is $\theta - \varpi$; we shall have

$$\delta\varpi = \pm\frac{nr\sin(\theta-\varpi)}{ae},$$

as the particle is moving to or from the nearer apse.

(6) A particle moving about a center of force in the focus, in an ellipse of small excentricity, receives a small impulse perpendicular to its direction of motion at any instant. Find the effect on the position of the apse.

(7) Again, if at the extremity of the axis minor the velocity be increased by $\frac{1}{n}$th, and the direction changed so

that h remains the same, find the alteration in the form and position of the orbit.

$$\delta a = 2\left(\frac{a^3}{\mu}\right)^{\frac{1}{2}} \delta V,$$

$$\delta e = \left(\frac{a}{\mu}\right)^{\frac{1}{2}} \left(\frac{1}{e} - e\right) \delta V.$$

(8) The first term of the central disturbing force on the moon is $-m^2r$, where the central force is $\frac{\mu}{r^2}$; shew that the apsidal angle (the orbit being nearly circular) is

$$\pi\left(1 + \frac{3}{2}\frac{m^2}{n^2}\right) \text{ nearly,}$$

where $\frac{2\pi}{n}$ is a mean lunar month.

(9) A particle is moving in a circle about a center of force $\propto$ (Dist.)$^{-2}$. The absolute force of the center increases slowly and uniformly. Determine the approximate elements of the orbit after a given time.

(10) A particle moves in a focal elliptic orbit in a very rare medium whose resistance is as the square of the velocity; determine the effect of the resistance on the periodic time.

(11) A particle is projected along a slightly rough inclined plane; find the approximate path, and the velocity at any point.

CHAPTER XII.

MOTION OF TWO OR MORE PARTICLES.

237. HAVING considered in detail the various cases which occur in the motion of a single particle subject to the action of any forces, and whose motion is either free, constrained or resisted, we proceed to the investigation of various cases in which more than one particle are involved. These will divide themselves naturally into two series; first, when the particles are entirely free, and are subject to their mutual attractions as well as other common impressed forces; and second, when there are in addition constraining forces; such as when two or more of the particles are connected by inextensible strings, &c. Let us take these in order:—

I. *Free Motion.*

238. It is an immediate deduction from the laws of motion that if two particles attract each other, they exert each on the other the same moving force.

For suppose them connected by a rigid rod without weight, and at rest; then it is evident that their mutual attraction will not set them in motion, but the pressure exerted by the rod is the same on each, and balances the moving forces of attraction, hence the latter are equal. If then m, m', be the masses of the particles, and the law of the force be $\phi'(D)$, we may express the common moving force by

$$mm'\phi'(D).$$

239. PROP. *A system of free particles is subject to no forces but their mutual attractions; to investigate the motion of the system.*

Let at time t, x_n, y_n, z_n be the co-ordinates in space of the particle whose mass is m_n, and let $\phi'(D)$ be the law of attraction. Let ${}_pr_q$ express the distance between the particles m_p and m_q; then we have for the motion of m_1,

$$m_1 \frac{d^2x_1}{dt^2} = \Sigma \left\{ m_1 m_n \phi'({}_1r_n) \frac{x_n - x_1}{{}_1r_n} \right\} \text{............} (1),$$

$$m_1 \frac{d^2y_1}{dt^2} = \Sigma \left\{ m_1 m_n \phi'({}_1r_n) \frac{y_n - y_1}{{}_1r_n} \right\} \text{............} (2),$$

$$m_1 \frac{d^2z_1}{dt^2} = \Sigma \left\{ m_1 m_n \phi'({}_1r_n) \frac{z_n - z_1}{{}_1r_n} \right\} \text{............} (3),$$

with similar equations for each of the others; the sums being taken throughout the system.

240. I. In the expression for $m_p \dfrac{d^2x_p}{dt^2}$, we have a term

$$m_p m_q \phi'({}_pr_q) \frac{x_q - x_p}{{}_pr_q},$$

and in $m_q \dfrac{d^2x_q}{dt^2}$ we have

$$m_q m_p \phi'({}_qr_p) \frac{x_p - x_q}{{}_qr_p}.$$

Hence if we add all the equations of the form (1) together the result will be

$$m_1 \frac{d^2x_1}{dt^2} + m_2 \frac{d^2x_2}{dt^2} + \text{......} = 0;$$

$$\text{or } \Sigma \left(m \frac{d^2x}{dt^2} \right) = 0.$$

Similarly

$$\Sigma \left(m \frac{d^2y}{dt^2} \right) = 0.$$

$$\Sigma \left(m \frac{d^2z}{dt^2} \right) = 0.$$

Now if $\bar{x}$, $\bar{y}$, $\bar{z}$, be at time t the co-ordinates of the center of gravity of all the particles,

$$\Sigma(m)\,\bar{x} = \Sigma(mx),$$
$$\Sigma(m)\,\bar{y} = \Sigma(my),$$
$$\Sigma(m)\,\bar{z} = \Sigma(mz).$$

And the above equations may be written,

$$\left.\begin{aligned} \Sigma(m)\frac{d^2\bar{x}}{dt^2} &= 0 \\ \Sigma(m)\frac{d^2\bar{y}}{dt^2} &= 0 \\ \Sigma(m)\frac{d^2\bar{z}}{dt^2} &= 0 \end{aligned}\right\} \quad \text{or,} \quad \begin{aligned} \frac{d^2\bar{x}}{dt^2} &= 0, \\ \frac{d^2\bar{y}}{dt^2} &= 0, \\ \frac{d^2\bar{z}}{dt^2} &= 0. \end{aligned}$$

Whence

$$\left.\begin{aligned} \frac{d\bar{x}}{dt} &= a \\ \frac{d\bar{y}}{dt} &= b \\ \frac{d\bar{z}}{dt} &= c \end{aligned}\right\}$$

which equations shew that the velocity of the center of gravity parallel to each of the co-ordinate axes remains invariable during the motion, that is, that *the center of gravity of the system remains at rest, or moves uniformly in a straight line.*

The values of a, b, c, may thus be determined,

$$a = \frac{d\bar{x}}{dt} = \frac{\Sigma\left(m\dfrac{dx}{dt}\right)}{\Sigma(m)}.$$

Now if the initial velocity of m_1 were resolvable into u_1, v_1, w_1, parallel to the axes respectively, and similarly for m_2, &c.

$$a = \frac{\Sigma(mu)}{\Sigma(m)}, \text{ and so for } b, \text{ \&c.}$$

241. II. Again, it is evident that if we multiply in succession equation (1) by y_1, and equation (2) by x_1, and subtract, and take the sum of all such remainders through the system of equations of the forms (1) and (2), we shall have

$$\Sigma m \left(y \frac{d^2x}{dt^2} - x \frac{d^2y}{dt^2}\right) = 0.$$

Or integrating once,

$$\Sigma m \left(y \frac{dx}{dt} - x \frac{dy}{dt}\right) = A_3.$$

Now if in the plane of xy we take ρ, θ the polar co-ordinates of the projection of m,

$$y \frac{dx}{dt} - x \frac{dy}{dt} = \rho^2 \frac{d\theta}{dt};$$

$$\therefore \Sigma m\rho^2 \frac{d\theta}{dt} = A_3.$$

Now if a_z be the area swept out by the radius vector ρ on the plane of xy,

$$\tfrac{1}{2}\rho^2 \frac{d\theta}{dt} = \frac{d(a_z)}{dt},$$

and our equation integrated gives

$$2\Sigma ma_z = A_3 t,$$

the constant being unnecessary if we agree to reckon a_z from the position of ρ at time $t = 0$.

This equation shews (since xy is *any* plane) that generally in the motion of a free system of particles subject only to their mutual attractions *the sum of the products of the mass of each particle of the system, and the area swept out by the radius vector of its projection on any plane, and about any point in that plane; will be proportional to the time.*

Take a_x, a_y to represent for the planes yz, xz the same that a_z represent for xy,

$$2\Sigma(ma_x) = A_1 t,$$

$$2\Sigma(ma_y) = A_2 t.$$

The value of this product for a plane, the direction cosines of whose normal are λ, μ, ν, will be

$$(\lambda A_1 + \mu A_2 + \nu A_3)\, t,$$

and will be a maximum if

$$\lambda A_1 + \mu A_2 + \nu A_3 \text{ is so,}$$

subject to the equation of condition

$$\lambda^2 + \mu^2 + \nu^2 = 1.$$

This gives $\lambda = \dfrac{A_1}{\sqrt{(A_1^2 + A_2^2 + A_3^2)}} = \dfrac{A_1}{A}$ suppose,

with similar values for μ and ν;

and the value of the product for the plane so found is evidently

$$At.$$

Hence, we see also, that *the plane for which the sum of the products of the masses of the particles into the sectorial areas described by the radii vectores of their projections is a maximum, is a fixed plane or parallel to a fixed plane during the motion.* It has been called on this account the *Invariable Plane.*

242. III. Multiply

$$(1) \text{ by } 2\frac{dx_1}{dt}, \quad (2) \text{ by } 2\frac{dy_1}{dt}, \quad (3) \text{ by } 2\frac{dz_1}{dt},$$

and treating similarly all the other equations, add them all together.

Let us consider the result as regards the term on the right-hand side involving the product $m_p m_q$.

Written at length it is

$$\frac{2m_p m_q \phi'({}_pr_q)}{{}_pr_q}\left\{(x_q - x_p)\frac{dx_p}{dt} + (x_p - x_q)\frac{dx_q}{dt}\right.$$

$$\left. + \text{ similar terms in } y \text{ and } z\right\};$$

and the portion in brackets is equal to

$$-\left\{(x_q - x_p)\frac{d}{dt}(x_q - x_p) + \text{similar terms is } y,\ z\right\};$$

$$\text{or, } -{}_pr_q\frac{d}{dt}({}_pr_q);$$

$$\therefore\ \Sigma\left\{2m\left(\frac{dx}{dt}\frac{d^2x}{dt^2} + \frac{dy}{dt}\frac{d^2y}{dt^2} + \frac{dz}{dt}\frac{d^2z}{dt^2}\right)\right\}$$

$$= -\Sigma\left\{2m_p m_q \phi'({}_pr_q)\frac{d}{dt}({}_pr_q)\right\};$$

or integrating,

$$\Sigma(mv^2) = C - 2\Sigma\{m_p m_q \phi({}_pr_q)\}.$$

And by taking this integral between limits, we see that—*the change in the Vis Viva of the system in any time depends only on the relative distances of the particles at the beginning and end of that time.*

243. So far for the case of several particles. The simplest examples will of course be found in the case of two particles only, and to such we will confine our attention, as when three or more are involved, the problem does not admit of exact solution, and in the two important applications which have been made of it, namely to the Lunar and Planetary Theories, it is found that a distinct method of approximation is required for each.

Since the acceleration of the center of gravity is zero, it follows that the motion of each particle with reference to that point is the same as if the latter were at rest. Also, since if we apply to each particle of the system an acceleration equal and opposite to that of any one of them, the latter will be reduced to rest, and therefore if there are only two, we see that the relative motion of one about the other will be the same as if the sum of the masses were substituted for the latter.

244. PROP. *Two particles moving initially with given velocities in the same straight line, are subject to no forces but their mutual attraction inversely as the square of the distance; to determine the motion.*

The motion will evidently be confined to the straight line. Let m, m' be the masses of the particles estimated on the hypothesis that unit of mass exerts unit of force at unit of distance; x, x' their distances at any time t from a fixed point in the line of motion, then

$$\left.\begin{aligned} m\frac{d^2x}{dt^2} &= \frac{mm'}{(x'-x)^2} \\ m'\frac{d^2x'}{dt^2} &= -\frac{mm'}{(x'-x)^2} \end{aligned}\right\} \text{..............} \quad (1).$$

Hence, if $\bar{x}$ be the co-ordinate of the center of gravity at time t,

$$m\frac{d^2x}{dt^2} + m'\frac{d^2x'}{dt^2} = (m+m')\frac{d^2\bar{x}}{dt^2} = 0,$$

$$m\frac{dx}{dt} + m'\frac{dx'}{dt} = (m+m')\frac{d\bar{x}}{dt} = C$$

$$= mV + m'V',$$

if V and V' be the initial velocities.

Integrating again,

$$mx + m'x' = (m+m')\,\bar{x} = (mV + m'V')\,t + C'$$

$$= (mV + m'V')\,t + ma + m'a' \text{.........} \quad (2),$$

if a, a' denote the initial positions of the particles.

Again, from equations (1),

$$\frac{d^2(x'-x)}{dt^2} = -\frac{m+m'}{(x'-x)^2},$$

from which, by multiplying by m or m', we see that the relative motion of the one with respect to the other, is the same as if the former had moved to the sum of the masses collected at the latter.

From the above $\left\{\frac{d(x'-x)}{dt}\right\}^2 = C + 2\frac{m+m'}{x'-x}$.

But $(V'-V)^2 = C + 2\frac{m+m'}{a'-a}$;

$$\therefore \left(\frac{d(x'-x)}{dt}\right)^2 = (V'-V)^2 + 2(m+m')\left\{\frac{1}{x'-x} - \frac{1}{a'-a}\right\} \dots (3).$$

This is of the form

$$\left(\frac{d\omega}{dt}\right)^2 = \frac{A}{\omega} \pm B;$$

$$\text{or, } t = \int \frac{\sqrt{(\omega)}\, d\omega}{\sqrt{(A \pm B\omega)}},$$

which may be integrated by putting $\omega = y^2$. The integral will be circular or logarithmic as B is $-$ or $+$. Thus we have $x'-x$ in terms of t, and knowing $mx + m'x'$ by (2), the motion is completely determined.

If at the instant of projection

$$(V-V')^2 = \frac{2(m+m')}{a'-a},$$

the formula (3) becomes

$$\sqrt{(x'-x)} \,.\, \frac{d(x'-x)}{dt} = \pm\sqrt{\{2(m+m')\}},$$

$$\tfrac{2}{3}(x'-x)^{\frac{3}{2}} = C \pm \sqrt{\{2(m+m')\}}\, t,$$

$$\tfrac{2}{3}(a'-a)^{\frac{3}{2}} = C,$$

and the motion is completely determined.

245. There is another method of treating this problem. Suppose instead of determining the relative motion of the particles, we consider that of each relatively to the common center of gravity. The distance of m from the center of gravity is

$$\bar{x} - x = \frac{mx + m'x'}{m + m'} - x = \frac{m'(x' - x)}{m + m'};$$

and we easily find from (1),

$$m'\left(\frac{d^2x'}{dt^2} - \frac{d^2x}{dt^2}\right) = -\frac{mm'}{(x' - x)^2} - \frac{m'^2}{(x' - x)^2};$$

$\therefore$ for the relative motion of m and the center of gravity,

$$m\frac{d^2(\bar{x} - x)}{dt^2} = -\frac{mm'}{(x' - x)^2}$$

$$= -\frac{mm'^3}{(m + m')^2(\bar{x} - x)^2};$$

whence $\bar{x} - x$ may be determined.

246. Prop. *Two particles anyhow projected, are acted on solely by their mutual attraction; to shew that the line joining them is always parallel to a fixed plane.*

If m and m' be the particles, x, y, z, x', y', z', their co-ordinates at time t, r their distance, and P the mutual moving force of attraction, we have the following equations,

$$m\frac{d^2x}{dt^2} = P\frac{x' - x}{r}, \quad m'\frac{d^2x'}{dt^2} = P\frac{x - x'}{r},$$

with similar expressions for the other co-ordinates; hence

$$\frac{\dfrac{d^2(x' - x)}{dt^2}}{x' - x} = \frac{\dfrac{d^2(y' - y)}{dt^2}}{y' - y} = \frac{\dfrac{d^2(z' - z)}{dt^2}}{z' - z},$$

and integrating,

$$(x' - x)\frac{d(y' - y)}{dt} - (y' - y)\frac{d(x' - x)}{dt} = C_3,$$

with other two similar equations. Therefore

$$C_3'(z'-z)+C_2(y'-y)+C_1(x'-x)=0.$$

Hence, the line joining the particles is always parallel to the plane whose direction-cosines are as C_1, C_2, C_3. This corresponds to Art. 241.

Also it is evident that the motion of the particles with respect to each other in a plane parallel to this is the same as if the plane were at rest (Art. 243).

From the preceding propositions the following are evident deductions.

The center of gravity of the two particles is at rest only when the initial velocities are zero, or when the directions of projection are the same or parallel, and the momenta equal and opposite.

The plane of relative motion will be at rest only when the initial directions lie in one plane.

If the force be inversely as the square of the distance, the relative orbits of the particles about each other, and therefore about their center of gravity, will be conic sections about a focus.

It is needless to pursue this any further, as the preceding results enable us to reduce the problem to cases treated of in former chapters.

247. PROP. *Two particles in space move under the action of given forces, as well as their mutual attraction; to determine the motion.*

Taking the same notation as in Art. 246, if X, Y, Z, X', Y', Z', be the resolved parts of the given forces, we have

$$m\frac{d^2x}{dt^2}=P\frac{x'-x}{r}+mX,\quad m'\frac{d^2x'}{dt^2}=P\frac{x-x'}{r}+m'X',$$

with similar equations for the other co-ordinates.

Hence, $$\frac{d^2(x'-x)}{dt^2}=-\frac{m+m'}{mm'}P\frac{x'-x}{r}+X'-X,$$

and so on. Thus we see that the relative motion about m will be found by applying to both particles reversed in direction the accelerating forces to whose action it is subjected.

Also, if $\bar{x}$, $\bar{y}$, $\bar{z}$, be the co-ordinates of the center of gravity, we find from the above three equations of the form

$$(m+m')\frac{d^2\bar{x}}{dt^2} = m\frac{d^2x}{dt^2} + m'\frac{d^2x'}{dt^2} = mX + m'X',$$

which shew that *the motion of the center of gravity is the same as if the particles had been collected into one, and acted on by the sum of the impressed moving forces.*

II. *Constrained Motion.*

248. In considering the constrained motion of particles, we can only take particular examples, but there are some general considerations which deserve attention.

If two particles be connected by an inelastic string, its only effect is to prevent their relative distance becoming greater than its own length. If we introduce then an unknown force T, for the tension of the string, the equations of motion can be written down, and the condition of the distance of the particles being equal to a given quantity will give us an additional equation, enabling us to eliminate, or to find the value of, the unknown force. If at any time the value of T become equal to zero, the motion of the particles must be investigated as if they were free, until the values of their co-ordinates shew that the string will begin to be tended again. In such a case, if their velocities resolved along the line joining them be not equal, an impact will take place, whose effects must be investigated by the methods of Chap. X.

In the case when the particles are connected by a rigid rod, we have exactly the same condition.

If there be more than two particles attached to the rod it may exert a transverse force; but cases of this kind more properly belonging to the Dynamics of a Rigid Body; and we therefore omit all consideration of them.

249. PROP. *Two particles attached to each other by an inextensible string, are projected with given velocities in space; to determine the motion.*

We may without loss of generality consider the distance between the particles at the instant of projection, to be equal to the length of the string. If their velocities are wholly perpendicular to its direction, or if their resolved parts along it are equal and in the same direction, there will be no impact. If not, suppose the masses m and m' to have velocities v and v' parallel to the string at the instant it is stretched. It is evident that the impact will change each of these into $\frac{mv' + m'v'}{m+m'}$. This then is determinate; so we may now in addition suppose the resolved parts of the velocities along the string equal to each other. Let x, y, z, x', y', z', be at any time the co-ordinates of the particles, then, if a be the length of the string,

$$m\frac{d^2x}{dt^2} = T\frac{x'-x}{a}, \quad m'\frac{d^2x'}{dt^2} = T\frac{x-x'}{a};$$

and so on.

Also, $$(x'-x)^2 + (y'-y)^2 + (z'-z)^2 = a^2$$

which are seven equations to find T, and the six co-ordinates of m and m'. From the form of the equations, or by treating them as in Art. 246, we see that the string remains parallel to a fixed plane, and that the center of gravity moves uniformly in a straight line; the motion of the particles about each other, and about it being the same as if it were at rest. Hence, the particles revolve with uniform angular velocity. From the above equations

$$T = \frac{mm'}{m+m'}\frac{V^2}{a},$$

where $V = \sqrt{\left[\left\{\frac{d(x'-x)}{dt}\right\}^2 + \left\{\frac{d(y'-y)}{dt}\right\}^2 + \left\{\frac{d(z'-z)}{dt}\right\}^2\right]}$,

is the relative velocity. The same result might have been easily obtained by considering that the velocity of m relative to the center of gravity is $\frac{m'V}{m+m'}$, that the radius of the circle it describes about that point is $\frac{m'a}{m+m'}$, and that T is the force which maintains it in that circle, and applying the formula in Art. 93.

250. PROP. *Two particles, connected by an inelastic string which passes over a small smooth pulley, move under the action of gravity; to determine the motion.*

This was partly anticipated in Art. 220. Let m, m' be the masses, and let x, x' denote their distances from the pulley at time t. Then if T be the tension of the string (the same throughout since the pulley is smooth), we have

$$m\frac{d^2x}{dt^2} = mg - T,$$

$$m'\frac{d^2x'}{dt^2} = m'g - T.$$

But $x + x' =$ length of string $= a$ suppose. Hence supposing $m > m'$,

$$(m + m')\frac{d^2x}{dt^2} = (m - m')\,g \quad \text{..............(1).}$$

This equation completely determines the motion. Also if we eliminate x and x', we have

$$T = \frac{2mm'}{m + m'}g,$$

and it is therefore constant.

This is one of the cases in which theoretical results may be tested by actual experiment with considerable accuracy. And it was this combination, with many delicate precautions against friction, &c. which Atwood made use of for experimental proof of the laws of motion.

We see for instance that by equation (1), we may easily keep $m + m'$ constant while $m - m'$ has any value, and thus by measuring the accelerations produced, find if they are in the same mass proportional to the moving forces. Again, keeping $m - m'$ constant, $m + m'$ may be varied at will. Hence by this process the third law of motion may be tested. Again, if while the weights are in motion a portion be suddenly removed from the greater so that they remain equal, (1) shews us that observation will enable us to test the first law of motion.

251. Instead of two masses connected by a string, suppose a flexible and uniform chain of length $2a$ hang over the pulley; then if x be the length hanging down on one side at time t, there will be $2a - x$ on the other, and the difference or

$$2(x-a),$$

is the portion whose weight accelerates the motion. Hence, $2\mu a$ being the mass of the chain, we have

$$2\mu a \frac{d^2x}{dt^2} = 2\mu g (x-a);$$

$$\therefore\ x - a = A\epsilon^{\sqrt{\left(\frac{g}{a}\right)}t} + B\epsilon^{-\sqrt{\left(\frac{g}{a}\right)}t}.$$

If the chain were initially at rest, a portion $a+b$ being on one side of the pulley,

$$b = A + B,$$

$$0 = A - B;$$

$$\therefore\ x - a = \frac{b}{2}\left\{\epsilon^{\sqrt{\left(\frac{g}{a}\right)}t} + \epsilon^{-\sqrt{\left(\frac{g}{a}\right)}t}\right\}.$$

This is true until $x = 2a$, that is, till the chain leaves the pulley; the value of t at that instant being t_0, we have

$$\frac{2a}{b} = \epsilon^{\sqrt{\left(\frac{g}{a}\right)}t_0} + \epsilon^{-\sqrt{\left(\frac{g}{a}\right)}t_0};$$

$$\therefore\ 2\sqrt{\left(\frac{a^2}{b^2} - 1\right)} = \epsilon^{\sqrt{\left(\frac{g}{a}\right)}t_0} - \epsilon^{-\sqrt{\left(\frac{g}{a}\right)}t_0},$$

whence the value of t_0 may be found.

If $b = \frac{3a}{5}$, i.e. if the portions of the chain were initially as 4 : 1,

$$t_0 = \sqrt{\left(\frac{a}{g}\right)}\log_e 3.$$

252. Prop. *A particle of mass m hangs over a small pulley, and the other end of the string is attached to a mass m′ lying on a smooth horizontal plane; to determine the motion.*

Let a be the length of the string, then it is evident that if b be the height of the pulley above the plane, θ the angle the string attached to m' makes with the vertical, and x the distance of m from the pulley, we must have

$$x + \frac{b}{\cos\theta} = a \quad\quad\quad (1).$$

Also if T be the tension of the string, the equations of motion are

$$\left.\begin{aligned} m\frac{d^2x}{dt^2} &= mg - T \\ m'\frac{d^2(b\tan\theta)}{dt^2} &= -T\sin\theta \end{aligned}\right\} \quad\quad (2).$$

These will suffice for the determination of the motion, and the tension of the string; unless $T\cos\theta$ should become greater than $m'g$, in which case the mass m' will be withdrawn from the plane, and instead of the last of equations (2) we must have two equations for the determination of its motion.

253. Prop. *Two particles m, m′, connected by a rigid rod, are forced to move one on each of two straight lines in a vertical plane, inclined at angles α, α′ to the vertical; to determine the motion, and the time of a small oscillation about the position of equilibrium.*

Let θ be the inclination of the rod to the vertical at time t, T its tension, R, R' the reactions of the lines, x, x' the distances of the particles from the point of intersection of the lines, and a the length of the rod.

Then,

$$\left.\begin{aligned} m\frac{d^2x}{dt^2} &= mg\cos\alpha + T\cos(\theta+\alpha) \\ m'\frac{d^2x'}{dt^2} &= m'g\cos\alpha' - T\cos(\theta-\alpha') \end{aligned}\right\} \quad\quad (1),$$

$$\left.\begin{aligned}R &= mg \sin\alpha + T\sin(\theta+\alpha)\\ R' &= m'g\sin\alpha' + T\sin(\theta-\alpha')\end{aligned}\right\}\dots\dots\dots (2),$$

$$\left.\begin{aligned}x &= x'\cos(\alpha+\alpha') - a\cos(\theta+\alpha)\\ x' &= x\cos(\alpha+\alpha') + a\cos(\theta-\alpha')\end{aligned}\right\}\dots\dots\dots (3).$$

These six equations give x, x', R, R', T and θ in terms of t, and thus theoretically complete the solution.

254. For the time of a small oscillation, we must first find the position of equilibrium. This will be obtained by equating to zero the right hand members of equations (1). Let $\bar{x}$, $\bar{x}'$, correspond to this position; and let

$$x = \bar{x} + \xi,\quad x' = \bar{x}' + \xi',$$

where ξ and ξ' are small.

Eliminating T from equations (1), and then θ by means of (3) we have, putting $\beta = \alpha + \alpha'$,

$$m\left(\frac{d^2x}{dt^2} - g\cos\alpha\right)(x' - x\cos\beta) - m'\left(\frac{d^2x'}{dt^2} - g\cos\alpha'\right)(x - x'\cos\beta) = 0.$$

But $x^2 + x'^2 - 2xx'\cos\beta = a^2$;

$\therefore$ neglecting quantities of the second order,

$$\bar{x}\xi + \bar{x}'\xi' - (\bar{x}\xi' + \bar{x}'\xi)\cos\beta = 0;$$

$$\text{or,}\quad \xi' = \xi\,\frac{\bar{x} - \bar{x}'\cos\beta}{\bar{x}\cos\beta - \bar{x}'}.$$

Hence eliminating ξ',

$$m\left(\frac{d^2\xi}{dt^2} - g\cos\alpha\right)\left\{\bar{x}' - \bar{x}\cos\beta + \xi\left(\frac{\bar{x} - \bar{x}'\cos\beta}{\bar{x}\cos\beta - \bar{x}'} - \cos\beta\right)\right\}$$

$$- m'\left(\frac{d^2\xi}{dt^2}\,\frac{\bar{x} - \bar{x}'\cos\beta}{\bar{x}\cos\beta - \bar{x}'} - g\cos\alpha'\right)$$

$$\left\{\bar{x} - \bar{x}'\cos\beta + \xi\left(1 - \frac{\bar{x} - \bar{x}'\cos\beta}{\bar{x}\cos\beta - \bar{x}'}\cos\beta\right)\right\} = 0.$$

Keeping only terms of the first order in ξ, since the terms not involving ξ or its differential cofficient must evidently vanish of themselves; we have

$$\{m\,(\bar{x}'-\bar{x}\cos\beta)^2+m'\,(\bar{x}-\bar{x}'\cos\beta)^2\}\,\frac{d^2\xi}{dt^2}$$
$$+g\sin^2\beta\,(m\bar{x}\cos\alpha+m'\bar{x}'\cos\alpha')\,\xi=0,$$

an equation of the form

$$\frac{d^2\xi}{dt^2}+n^2\xi=0;$$

and the time of oscillation is $\frac{2\pi}{n}$, as may readily be found.

255. Prop. *Two particles m and m' are attached at different points to an inelastic string, one of whose extremities is fixed. If the system be displaced, to determine the motion.*

Take the axes of x and y horizontal, and that of z vertically downwards, the extremity of the string being origin.

Let a, a' be the lengths of the portions of the string, θ, θ' the angles they make with the vertical, ϕ, ϕ' the angles which vertical planes through them at time t, make with the plane of xz. Let x, y, z, x', y', z', be the co-ordinates of the particles and T, T' the tensions of the strings.

Then

$$\left.\begin{aligned} m\frac{d^2x}{dt^2} &= -\,T\sin\theta\cos\phi+T'\sin\theta'\cos\phi',\\ m\frac{d^2y}{dt^2} &= -\,T\sin\theta\sin\phi+T'\sin\theta'\sin\phi',\\ m\frac{d^2z}{dt^2} &= mg-T\cos\theta+T'\cos\theta', \end{aligned}\right\}$$

$$\left.\begin{aligned} m'\frac{d^2x'}{dt^2} &= -\,T\sin\theta'\cos\phi',\\ m'\frac{d^2y'}{dt^2} &= -\,T'\sin\theta'\sin\phi',\\ m'\frac{d^2z'}{dt^2} &= m'g-T'\cos\theta'. \end{aligned}\right\}$$

Besides these, we have the six equations for x, y, z, x', y', z' in terms of a, a', θ, ϕ, θ', ϕ', in all twelve equations for the determination of the twelve unknowns in terms of t.

256. These equations will be much simplified if we consider the displacement to be in one plane, as the motion will evidently be confined to that plane. By this means we at once get rid of y, y', ϕ and ϕ'. A still greater simplification will be obtained by taking in addition the condition that θ and θ' are so small, that their squares and higher powers may be neglected. With these our equations become

$$\left.\begin{aligned} m\frac{d^2x}{dt^2} &= -T\theta + T'\theta', \\ m\frac{d^2z}{dt^2} &= mg - T + T', \end{aligned}\right\}$$

$$\left.\begin{aligned} m'\frac{d^2x'}{dt^2} &= -T'\theta', \\ m'\frac{d^2z'}{dt^2} &= -m'g - T'. \end{aligned}\right\}$$

And
$$x = a\theta,$$
$$x' = a\theta + a'\theta',$$
$$z = a,$$
$$z' = a + a'.$$

Hence, $T' = m'g$, and $T = (m + m')g$,

$$\left.\begin{aligned} ma\frac{d^2\theta}{dt^2} &= -(m + m')g\theta + m'g\theta', \\ m'\left(a\frac{d^2\theta}{dt^2} + a'\frac{d^2\theta'}{dt^2}\right) &= -m'g\theta'. \end{aligned}\right\}$$

Introducing an indeterminate multiplier, and adding,

$$(m + \lambda m')\frac{d^2\theta}{dt^2} + \lambda m'\frac{a'}{a}\frac{d^2\theta'}{dt^2} + \frac{g}{a}\{(m+m')\theta + m'(\lambda - 1)\theta'\} = 0.$$

Let λ_1, λ_2 be the roots of the equation

$$\frac{\lambda}{m+\lambda m'} \frac{a'}{a} = \frac{\lambda - 1}{m+m'}.$$

Evidently one is positive and the other negative, and the form of the equation shews that for both $m+\lambda m'$ is positive.

Put $\phi = \theta + \dfrac{\lambda m'}{m+\lambda m'} \dfrac{a'}{a} \theta' = \theta + k\theta'$, suppose.

Then the above equation gives

$$\frac{d^2\phi}{dt^2} + \frac{g}{a} \cdot \frac{m+m'}{m+\lambda m'} \phi = 0.$$

By the recent remark the coefficient of ϕ is positive for both values of λ; let its values be n_1^2 and n_2^2, and we have, k_1, ϕ_1, k_2, ϕ_2, being the corresponding values of k and ϕ,

$$\phi_1 = \theta + k_1\theta' = \alpha_1 \cos(n_1 t + \beta_1),$$

$$\phi_2 = \theta + k_2\theta' = \alpha_2 \cos(n_2 t + \beta_2),$$

where α_1, α_2, β_1, β_2, are arbitrary constants.

Hence,

$$\theta = \frac{1}{k_2 - k_1} \{k_2 \alpha_1 \cos(n_1 t + \beta_1) - k_1 \alpha_2 \cos(n_2 t + \beta_2)\},$$

$$\theta' = \frac{1}{k_1 - k_2} \{\alpha_1 \cos(n_1 t + \beta_1) - \alpha_2 \cos(n_2 t + \beta_2)\}.$$

Having given the initial values of θ, θ', $\dfrac{d\theta}{dt}$ and $\dfrac{d\theta'}{dt}$, we find α_1, α_2, β_1, β_2, and thus the solution is complete. It may be noticed that the values of θ and θ' may be found at any time by taking the algebraic sum of the corresponding values of the inclinations to the vertical of two pendulums whose times of oscillation are $\dfrac{2\pi}{n_1}$ and $\dfrac{2\pi}{n_2}$. Also, if n_1, n_2 be commensurable, the system will in time return to its first position, and the motion will be periodic. This is a simple example of the principle of the "Coexistence of Small Oscillations;" but this, for its satisfactory treatment in the general case requires the use of D'Alembert's Principle, which is beyond the professed limits of the present treatise.

257. The problems which have just been given, may suffice to convey an idea of the mode of applying our methods to any proposed case of motion of two constrained particles. These methods are applicable to more complicated cases, when more particles than two are involved; but nothing would be gained by such a proceeding, as D'Alembert's Principle, above noticed, supplies us with a far simpler mode of investigating the motions of any system of free or connected particles.

EXAMPLES.

(1) Two spheres whose masses are M and M' are placed in contact, and one of them is projected in the line of centers with velocity V. If the law of attraction be D^{-2}, find where, and after what time, they will meet.

(2) If the sun were broken up into an indefinite number of fragments, uniformly filling the sphere of which the earth's orbit is a great circle, shew that each would revolve in a year.

(3) A sphere of mass M is driven out symmetrically by an internal explosion. Shew that if when the shell has a radius a, the outward velocity of each particle be v, the fragments can never be collected by their mutual attraction unless

$$v^2 < \frac{M}{a}.$$

(4) Two equal particles are initially at rest in two smooth tubes at right angles to each other. Shew that whatever were their positions, and whatever their law of attraction, they will reach the intersection of the tubes together.

(5) In last question suppose the original distances from the intersection of the tubes to be a, b, and the attraction as

the square of the distance inversely, find the future paths if at any instant the constraint is removed.

(6) A number of equal particles, attracting each other directly as the distance, are constrained to move in parallel tubes; if the positions of the particles be given at the commencement of the motion, determine the subsequent motion of each; and shew that the particles will oscillate symmetrically with respect to the plane perpendicular to the tubes which passed through their center of gravity at the commencement of the motion.

(7) Two given masses are connected by a slightly elastic string, and projected so as to whirl round, find the time of a small oscillation in the length of the string.

Give a numerical result, supposing the masses to weigh 1 lb. and 2 lbs. respectively, and the natural length of the string to be 1 yard, and supposing that it stretches $\frac{1}{10}$th inch for a tension of 1 lb.

(8) Two equal masses M, are connected by a string which passes through a hole in a smooth horizontal plane. One of them hanging vertically, shew that the other describes on the plane a curve whose differential equation is

$$\frac{d^2u}{d\theta^2} + \frac{u}{2} - \frac{g}{2h^2u^2} = 0,$$

and that the tension of the string is

$$M\frac{g + h^2u^3}{2}.$$

(9) Two equal particles connected by a string are placed in a circular tube. In the circumference is a center of force $\propto \frac{1}{D}$. One particle is initially at its greatest distance from the center of force, shew that if v, v' be the velocities with which they pass through a point 90° from the center of force,

$$\epsilon^{-\frac{v^2}{\mu}} + \epsilon^{-\frac{v'^2}{\mu}} = 1.$$

(10) Two equal balls repelling each other with a force $\propto \frac{1}{D^2}$ hang from the same point by strings of length l. Shew that if when in equilibrium, the strings making an angle 2α with each other, they be approximated by equal small arcs, the time of an oscillation is the same as that of a pendulum whose length is

$$\frac{l \cos \alpha}{1 + 2 \cos^2 \alpha}.$$

(11) One of two equal particles connected by an inelastic string moves in a straight groove. The other is projected parallel to the groove, the string being stretched; determine the motion, and shew that the greatest tension is four times the least.

(12) Two particles are connected by an elastic string of length $2a$, and one is projected perpendicularly to the string when it is unstretched. Shew that in the relative orbit

$$\frac{1}{a^2} - \frac{1}{p^2} \propto (r - a)^2.$$

(13) Two equal particles connected by a rigid rod move on a vertical circle. If they be slightly disturbed from the higher position of equilibrium, determine the motion.

Also find the time of a small oscillation about the position of stable equilibrium.

(14) Two particles P and Q are connected by a rigid rod. P is constrained to move in a smooth horizontal groove. If the particles be initially at rest, PQ making a given angle with the groove in a vertical plane through it, find the velocity of Q when it reaches the groove, and shew that Q's path in the vertical plane is an ellipse.

(15) A particle of mass m has attached to it two equal weights m', by means of strings passing over pulleys in the same horizontal plane, and is initially at rest halfway between them. Determine the motion. Shew that if the distance

between the pulleys be $2a$, the velocity of m will be zero when it has fallen through a space

$$\frac{4mm'a}{4m'^2 - m^2},$$

and that it will have its greatest velocity when passing through its position of statical equilibrium.

(16) Two masses M, M' are connected by a string which passes over a smooth peg. To M' is attached a string which supports a mass m such that $M' + m = M$, and m is displaced through an angle α. Investigate the motion, supposing m so small that the horizontal motion of M' may be neglected. Shew that the string $M'm$ will be vertical after the time

$$\left(\frac{\lambda}{g}\right)^{\frac{1}{2}} \int_0^\alpha \left(\frac{1 - \frac{m}{2M}\sin^2\theta}{\cos\theta - \cos\alpha}\right)^{\frac{1}{2}} d\theta,$$

where λ is the length of $M'm$.

APPENDIX.

A. Prop. *To find the time of fall from rest down any arc of an inverted cycloid.*

Let O be the point from which the particle commences its motion. Draw OA' parallel to CA, and on BA' describe

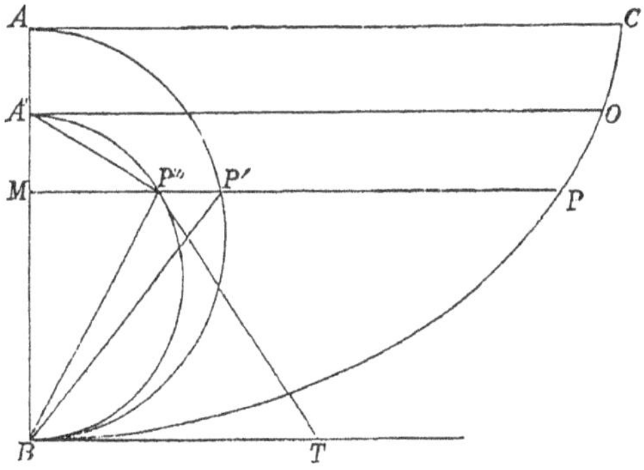

a semicircle. Let P, P', P'' be corresponding points of the curve, the generating circle, and the circle just drawn, and let us compare the velocities of the particle at P, and the point P''. Let $P''T$ be the tangent at P''.

$$\frac{\text{velocity of } P''}{\text{velocity of } P} = \frac{\text{element at } P''}{\text{element at } P}$$

$$= \frac{P''T}{BP'} = \frac{P''T}{BP''}\sqrt{\left(\frac{A'B}{AB}\right)}$$

$$= \frac{A'B}{2A'P''}\sqrt{\left(\frac{A'B}{AB}\right)}.$$

But velocity of $P = \sqrt{(2g\,.\,A'M)} = \sqrt{\left(\frac{2g}{A'B}\right)}\,.\,A'P''$.

Hence velocity of $P'' = \sqrt{\left(\frac{g}{2AB}\right)}.A'B$, a constant.

And as the length of $A'P''B$ is $\frac{\pi}{2}.A'B$,

time from A' to B in circle = time from O to B in cycloid

$$= \pi\sqrt{\left(\frac{AB}{2g}\right)}.$$

COR. It is evident from the proof, that the particle descends half the vertical space to B in half the time it takes to reach B.

B. PROP. *To find the nature of the brachistochrone, gravity being the only impressed force.*

The following is founded on Bernoulli's original solution. (WOODHOUSE, *Isoperimetrical Problems.*)

From Art. 140 it is evident that the curve lies in the vertical plane which contains the given points. Also it is easy to see that if the time of descent through the entire curve is a minimum, that through any portion of the curve is less than if that portion were changed into any other curve.

Also when a quantity is a maximum or minimum, since its first differential coefficient vanishes, contiguous values vary

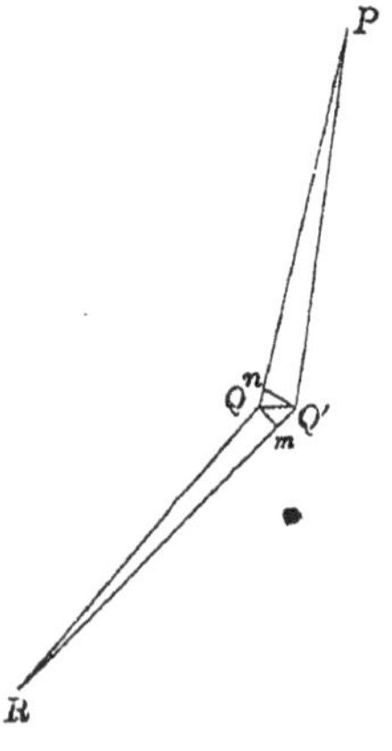

very slowly; so that if PQ, QR be two elements of the brachistochrone considered as a polygon, and if Q' be taken in the same horizontal line with Q and indefinitely near to it, the time down PQ', $Q'R$ is to the first order of small quantities the same as that down PQ, QR. Hence if v be the velocity down PQ (supposed uniform) and v' that down QR, drawing Qm, $Q'n$ perpendicular to RQ', PQ, we must have

$$\frac{Qn}{v} = \frac{Q'm}{v'}.$$

Now if θ be the inclination of PQ to the horizon, θ' that of QR, $Qn = QQ' \cos\theta$, $Q'm = QQ' \cos\theta'$, and

$$\therefore \frac{\cos\theta}{v} = \frac{\cos\theta'}{v'}.$$

This is true for any two consecutive elements of the required curve; therefore throughout the curve

$$v \propto \cos\theta.$$

But $v^2 \propto$ vertical space fallen through. Hence the curve required is such that the cosine of the angle it makes with the horizontal line through the point of departure varies as the square root of the distance from that line, which is easily seen to be a property of the cycloid if we remember that the tangent to that curve is parallel to the corresponding chord of its generating circle. For in the fig. p. 138,

$$\cos OP'N = \cos OAP' = \frac{AP'}{AO} = \sqrt{\left(\frac{AN}{AO}\right)} \propto \sqrt{(AN)}.$$

The brachistochrone then, in the case of gravity being the only impressed force, is an inverted cycloid whose cusp is at the point from which the particle descends.

B_1. Were there any number of impressed forces we might suppose their resultant constant in magnitude and direction for two successive elements. Then reasoning similar to that in Art. 140 would shew that the osculating plane to the brachistochrone always contains the resultant force. Again we should have as in last Article, $\frac{\cos\theta}{v} = \frac{\cos\theta'}{v'}$ where θ is now the complement of the angle between the curve and the resultant of the impressed forces.

Let that resultant $= F$, and let the element $PQ = \delta s$, and $\theta' = \theta + \delta\theta$. Then since F is supposed the same at P and Q,

$$v'^2 - v^2 = 2F\,\delta s \sin\theta \text{ (by Chap. IV.)},$$

$$\text{or } v\delta v = F\delta s \sin\theta.$$

But $v \propto \cos\theta$;

$$\therefore \frac{\delta v}{v} = -\frac{\sin\theta}{\cos\theta}\,\delta\theta.$$

Hence
$$\frac{v^2}{\frac{\delta s}{\delta\theta}} = -F\cos\theta.$$

But in the limit $\frac{\delta s}{\delta\theta} = \rho$, the radius of absolute curvature at Q, and $F\cos\theta$ is the resolved impressed force perpendicular to the curve. Hence in the general brachistochrone the pressure due to centrifugal force equals that due to the impressed forces, our result of Art. 147.

B_2. Now for the unconstrained path from P to R we have $\int v ds$ a minimum. Hence in the same way as before, ϕ being the angle corresponding to θ, $v\cos\phi = v'\cos\phi'$ from element to element, and therefore throughout the curve.

But in the brachistochrone,

$$\frac{\cos\theta}{v} = \frac{\cos\theta'}{v'}.$$

Now if the velocities in the two paths be equal at any surface of equilibrium, they will be equal in every other. Hence taking the angles for any surface of equilibrium

$$\cos\theta\cos\phi = \text{constant}.$$

As an example, suppose a parabola with its vertex upwards to have for directrix the base of an inverted cycloid; these curves evidently satisfy the above condition, the one being the free path, the other the brachistochrone, for gravity, and the velocities being in each due to the same horizontal line. And it is seen at once that the product of the cosines of the angles which they make with any horizontal straight line which cuts both is a constant whose magnitude depends on that of the cycloid and parabola, its value being $\sqrt{\frac{l}{4a}}$ where l is the latus rectum of the parabola, and a the diameter of the generating circle of the cycloid.

C. Prop. *To shew that of two curves both concave in the sense of gravity, joining the same points in a vertical plane and not meeting in any other point, a particle will descend the enveloped in less time than it will the enveloping curve; the initial velocity being the same in both cases.*

Take the axis of x as the line to a fall from which the initial velocity is due, and the axis of y in the sense of gravity, then

$$\frac{ds}{dt} = \sqrt{(2gy)};$$

$$\therefore\ t_1 = \int_{x_1}^{x_2} \frac{\sqrt{\left\{1+\left(\frac{dy}{dx}\right)^2\right\}}\,dx}{\sqrt{(2gy)}}$$

$$\propto \int_{x_1}^{x_2} \frac{\sqrt{(1+p^2)}}{\sqrt{(y)}}\,dx \ldots\ldots\ldots\ldots\ldots\ldots(1);$$

$$\therefore\ \delta t_1 \propto \int_{x_1}^{x_2} dx\,\delta y\left(N - \frac{dP}{dx}\right),$$

(since the limits are constant),

$$\propto \int_{x_1}^{x_2} \delta y dx \left[-\tfrac{1}{2}\frac{\sqrt{(1+p^2)}}{y\sqrt{(y)}} + \tfrac{1}{2}\frac{p^2}{y\sqrt{(y)}\sqrt{(1+p^2)}} - \frac{q}{\sqrt{(y)}\sqrt{(1+p^2)}} + \frac{p^2 q}{\sqrt{(y)}\,(1+p^2)^{\frac{3}{2}}}\right]$$

$$\propto \int_{x_1}^{x_2} \delta y\,dx\,\frac{-(1+p^2)+2p^2qy-2qy\,(1+p^2)}{y\sqrt{(y)}\,(1+p^2)^{\frac{3}{2}}}$$

$$\propto \int_{x_1}^{x_2} \delta y\,dx\,\frac{-(1+p^2+2yq)}{y\sqrt{(y)}\,(1+p^2)^{\frac{3}{2}}}.$$

Now the curve is convex to the axis of x, hence yq is positive, and by (1) $\sqrt{(y)}$ and $\sqrt{(1+p^2)}$ have the same sign. Hence the sign of δt_1 is the opposite of that of δy, and for an enveloping curve δy is negative. Hence the time of fall will be longer.

We may thus pass from one curve to any other enveloping one, even situated at a finite distance, provided the latter be concave throughout; else the multiplier of $\delta y . dx$ in the integral might change sign between the limits. (BERTRAND. *Liouville's Journal*, Vol. VII.)

D. PROP. *To find the curve in which the time of descent to the lowest point is a given function $\phi(a)$ of a, the vertical height fallen through.*

$$\text{Here } \sqrt{(2g)}\, t = \phi(a) = \int_0^a \frac{ds}{\sqrt{(a-x)}}.$$

From this equation s is to be found as a function of x. Let us consider the integral

$$\int_0^1 y^{\alpha-1}(1-y)^{\beta-1}\, dy = \frac{\Gamma(\alpha)\,\Gamma(\beta)}{\Gamma(\alpha+\beta)}.$$

(GREGORY'S *Exam.* p. 471.)

Put $\beta = \frac{1}{2}$, we have

$$\int_0^1 \frac{y^{\alpha-1}\, dy}{(1-y)^{\frac{1}{2}}} = \frac{\Gamma(\alpha)\,\Gamma(\frac{1}{2})}{\Gamma(\alpha+\frac{1}{2})} = \sqrt{(\pi)}\,\frac{\Gamma(\alpha)}{\Gamma(\alpha+\frac{1}{2})}.$$

Put $z = ay$, then the limits are 0, a,

$$\int_0^a \frac{\left(\frac{z}{a}\right)^{\alpha-1} \frac{dz}{a}}{\left(1-\frac{z}{a}\right)^{\frac{1}{2}}} = \sqrt{(\pi)}\,\frac{\Gamma(\alpha)}{\Gamma(\alpha+\frac{1}{2})};$$

$$\therefore \int_0^a \frac{z^{\alpha-1}\, dz}{(a-z)^{\frac{1}{2}}} = \sqrt{(\pi)}\, a^{\alpha-\frac{1}{2}}\,\frac{\Gamma(\alpha)}{\Gamma(\alpha+\frac{1}{2})}.$$

$$\therefore \int_0^x \frac{da}{(x-a)^{\frac{1}{2}}} \int_0^a \frac{z^{\alpha-1}\, dz}{(a-z)^{\frac{1}{2}}} = \sqrt{(\pi)}\,\frac{\Gamma(\alpha)}{\Gamma(\alpha+\frac{1}{2})} \int_0^x \frac{a^{\alpha-\frac{1}{2}}\, da}{(x-a)^{\frac{1}{2}}}.$$

Put $a = xy$,

$$\int_0^x \frac{a^{\alpha-\frac{1}{2}}\,da}{(x-a)^{\frac{1}{2}}} = \int_0^1 \frac{x^{\alpha+\frac{1}{2}}\,y^{\alpha-\frac{1}{2}}\,dy}{x^{\frac{1}{2}}\,(1-y)^{\frac{1}{2}}}$$

$$= x^\alpha \int_0^1 \frac{y^{\alpha-\frac{1}{2}}\,dy}{(1-y)^{\frac{1}{2}}}$$

$$= x^\alpha\,\frac{\Gamma(\alpha+\frac{1}{2})\,\Gamma(\frac{1}{2})}{\Gamma(\alpha+1)}.$$

Now $\Gamma(\alpha+1) = \alpha\Gamma(\alpha)$;

$$\therefore \int_0^x \frac{da}{(x-a)^{\frac{1}{2}}} \int_0^a \frac{z^{\alpha-1}\,dz}{(a-z)^{\frac{1}{2}}} = \frac{\pi}{\alpha}\,x^\alpha.$$

Multiply by $\alpha\phi(\alpha)\,d\alpha$, and integrate with respect to α,

$$\therefore \int_0^x \frac{da}{(x-a)^{\frac{1}{2}}} \int_0^a \frac{\{\int \alpha\phi(\alpha)\,z^{\alpha-1}\,d\alpha\}\,dz}{(a-z)^{\frac{1}{2}}} = \pi \int \phi(\alpha)\,x^\alpha\,d\alpha.$$

Let $$\int \phi(\alpha)\,x^\alpha\,d\alpha = f(x);$$

$$\therefore \int \alpha\phi(\alpha)\,x^{\alpha-1}\,d\alpha = f'(x)$$

$$\text{and } \int \alpha\phi(\alpha)\,z^{\alpha-1}\,d\alpha = f'(z),$$

and our equation becomes

$$\int_0^x \frac{da}{(x-a)^{\frac{1}{2}}} \int_0^a \frac{f'(z)\,dz}{(a-z)^{\frac{1}{2}}} = \pi f(x) \;\ldots\ldots\; (1).$$

Now $\phi(a) = \int_0^a \frac{ds}{(a-x)^{\frac{1}{2}}}$.

$$\therefore \quad \frac{1}{\pi}\int_0^x \frac{\phi(a)\,da}{(x-a)^{\frac{1}{2}}} = \frac{1}{\pi}\int_0^x \frac{da}{(x-a)^{\frac{1}{2}}} \int_0^a \frac{ds}{(a-z)^{\frac{1}{2}}}$$

$$= s \text{ (by (1))}.$$

$$\therefore \quad s = \frac{1}{\pi}\int_0^x \frac{\phi(a)\,da}{(x-a)^{\frac{1}{2}}},$$

which gives the equation to the required curve. (ABEL, *Œuvres*, Tom. I.)

Ex. I. Suppose the Tautochrone be required

$$\phi(a) = \sqrt{(2g)}\, t_0;$$

$$\therefore \quad s = \frac{\sqrt{(29)}t_0}{\pi}\int_0^x \frac{da}{(x-a)^{\frac{1}{2}}} = \frac{2\sqrt{(2g)}\, t_0}{\pi}\, x^{\frac{1}{2}};$$

$$\text{or} \qquad s^2 = \frac{8gt_0^2}{\pi^2}\, x; \text{ a cycloid.}$$

Ex. II. Let $\phi(a) = \sqrt{(2g)}\,\dfrac{a}{c}$, or the time proportional to the vertical height fallen through.

$$\therefore \quad \frac{cs}{\sqrt{(2g)}} = \int_0^x \frac{ada}{(x-a)^{\frac{1}{2}}} = \sqrt{(\pi)}\, x^{\frac{3}{2}}\frac{\Gamma(2)}{\Gamma(2+\frac{1}{2})} = \frac{4}{3}x^{\frac{3}{2}},$$ the equation to the required curve.

THE END.

MATHEMATICAL CLASS BOOKS,

FOR

COLLEGES AND SCHOOLS.

Mr. Todhunter's Algebra for Beginners. [*Preparing.*

Prof. Wilson's Treatise on Dynamics. 9*s.* 6*d.*

Cambridge Senate-House Problems, 1848 *to* 1851.
Solutions by Messrs. FERRERS and JACKSON. 15*s.* 6*d.*

Cambridge Senate-House Riders, 1848 *to* 1851.
Solutions by Mr. JAMESON. 7*s.* 6*d.*

Cambridge Senate-House Problems, Jan. 8, 1852.
Solutions. 1*s.* 6*d.*

Cambridge Senate-House Problems and Riders, 1854.
Solutions by the Moderators, Messrs. WALTON and MACKENZIE, 10*s.* 6*d.*

GREEK AND LATIN CLASS BOOKS.

Mr. Drake's Eumenides of Æschylus. With English Notes. 7*s.* 6*d.*

Mr. Drake's Demosthenes de Corona. With English Notes. 5*s.*

Mr. Frost's Thucydides, Book VI. With English Notes. 7*s.* 6*d.*

Dr. Humphreys' Exercitationes Iambicæ. 5*s.* 6*d.*

Mr. Mayor's Juvenal. With English Notes. 10*s.* 6*d.*

Mr. Merivale's Sallust. With English Notes. 5*s.*

Mr. Thring's Construing Book. 2*s.* 6*d.*

Mr. Wright's Hellenica. With English Notes. 3*s.* 6*d.*

Mr. Wright's Help to Latin Grammar. 4*s.* 6*d.*

ENGLISH GRAMMARS.

Mr. Thring's Elements of Grammar. 2*s.*

Mr. Thring's Child's Grammar. 1*s.*

Mr. Parminter's Materials for English Grammar. 3*s.* 6*d.*

PROSPECTUS of a Series of **Manuals for Theological Students** now in course of publication by MACMILLAN and Co., Cambridge.

IT is now upwards of three years since the Prospectus of this series was first issued. Three volumes have now been published, and several others are in an advanced state. The reception which the volumes already published have met with has fully justified the anticipation with which the publishers commenced the series, and warrants them in the belief, that their aim of supplying books "concise, comprehensive, and accurate," "convenient for the professional Student and interesting to the general reader," has been not unsuccessfully fulfilled.

The following paragraphs appeared in the original Prospectus, and may be here conveniently reproduced:—

"The Authors being Clergymen of the English Church, and the series being designed primarily for the use of Candidates for office in her Ministry, the books will seek to be in accordance with her spirit and principles; and therefore in treating of the opinions and principles of other communions, every effort will be made to avoid acrimony or misrepresentation.

"It will be the aim of the writers throughout the series to avoid all dogmatic expression of doubtful or individual opinions."

THE FOLLOWING ARE NOW READY.

I. *A History of the Christian Church from the* SEVENTH CENTURY TO THE REFORMATION. By the Rev. CHARLES HARDWICK, M.A., Fellow of St Catharine's Hall, Divinity Lecturer of King's College, and Christian Advocate in the University of Cambridge, Author of "A History of the XXXIX Articles." *With Four Maps constructed for this Work by A. Keith Johnston.* Crown 8vo. cloth, 10*s.* 6*d.*

OPINIONS OF THE PRESS.

"*As a manual for the student of Ecclesiastical History in the Middle Ages, we know no English work which can be compared to* Mr HARDWICK'S *book. It has two great merits, that it constantly refers the reader to the authorities, both original and critical, on which its statements are founded; and that it preserves a just proportion in dealing with various subjects.*"—GUARDIAN, April 12, 1854.

Mr Hardwick's *Middle-Age Church History.*

Opinions of the Press—*continued.*

"*This forms one of a series of Theological Manuals which* Messrs Macmillan, *Cambridge, have in course of publication. If the other volumes of the series are as well and carefully written as this, theological students will have good cause to thank them.*"—Clerical Journal, Sept. 22, 1853.

"*It is full in references and authority, systematic and formal in division, with enough of life in the style to counteract the dryness inseparable from its brevity, and exhibiting the results rather than the principles of investigation.* Mr Hardwick *is to be congratulated on the successful achievement of a difficult task.*"—Christian Remembrancer, October, 1853.

"*He has bestowed patient and extensive reading on the collection of his materials; he has selected them with judgment; and he presents them in an equable and compact style.*"—Spectator, September 17, 1853.

"*This book is one of a promised series of* 'Theological Manuals.' *In one respect, it may be taken as a sign of the times. It is a small unpretending volume in appearance, but it is based on learning enough to have sufficed, half a century since, for the ground of two or three quartos, or at least for several portly octavos. For its purpose it is admirable, giving you a careful and intelligent summary of events, and at the same time indicating the best sources of information for the further guidance of the student. Among the authorities thus referred to, we find the most modern as well as the most ancient, the Continental as well as the English.*"—British Quarterly, Nov. 1853.

"*It is distinguished by the same diligent research and conscientious acknowledgment of authorities which procured for* Mr Hardwick's 'History of the Articles of Religion' *such a favourable reception.*"—Notes and Queries, October 8, 1853.

"*To a good method and good materials* Mr Hardwick *adds that great virtue, a perfectly transparent style. We did not expect to find great literary qualities in such a manual, but we* have *found them: we should be satisfied in this respect with conciseness and intelligibility; but while this book has both, it is also elegant, highly finished, and highly interesting.*"—Nonconformist, November 30, 1853.

II. *A History of the Book of Common Prayer,*

together with a Rationale of the several Offices. By the Rev. FRANCIS PROCTER, M.A., Vicar of Witton, Norfolk, and late Fellow of St Catharine's Hall. Crown 8vo. cloth, 10*s*. 6*d*.

Opinions of the Press.

"Mr Procter's 'History of the Book of Common Prayer' *is by far the best commentary extant......Not only do the present illustrations embrace the whole range of original sources indicated by* Mr Palmer, *but* Mr Procter *compares the present Book of Common Prayer with the Scotch and American forms: and he frequently sets out in full the Sarum Offices. As a manual of extensive information, historical and ritual, imbued with sound Church principles, we are entirely satisfied with* Mr Procter's *important volume.*"—Christian Remembrancer, April, 1855.

Mr PROCTER, *on the Book of Common Prayer.*

OPINIONS OF THE PRESS—*continued.*

"*It is a résumé of all that has been done in the way of investigation in reference to the Prayer-Book.*"—ATHENÆUM, Feb. 17, 1855.

"*We can have little doubt that* Mr PROCTER's *History of our Liturgy will soon supersede the well-known work of* Wheatley, *and become a much-used hand-book beyond the circuits of the University for the more immediate use of which it has been produced.*"—NOTES AND QUERIES, March, 1855.

"*Although very decidedly anti-Roman in its tone, we gladly accept it as a substitute for the dull and dreary dogmatism of* Wheatley. *It presents in a popular and agreeable narrative, the history of those variations to which so much attention has been directed during the late eventful controversies; and while it contains a very careful, learned and scholarlike exposition of these changes, it also furnishes a most valuable commentary on the successive texts of the formularies themselves, as they are exhibited either in the original editions, or in the useful manuals of* Bulley *and* Keeling."—DUBLIN REVIEW (*Roman Catholic*), April, 1855.

"*We can speak with just praise of this compendious but comprehensive volume. It appears to be compiled with great care and judgment, and has profited largely by the accumulated materials collected by the learning and research of the last fifty years. It is a manual of great value to the student of Ecclesiastical History and of almost equal interest to every admirer of the Liturgy and Services of the English Church.*"—LONDON QUARTERLY REVIEW, April, 1855.

"*It is indeed a complete and fairly-written history of the Liturgy; and from the dispassionate way in which disputed points are touched on, will prove to many troubled consciences what ought to be known to them, viz.:—that they may without fear of compromising the principles of evangelical truth, give their assent and consent to the contents of the Book of Common Prayer.* Mr PROCTER *has done a great service to the Church by this admirable digest.*"—CHURCH OF ENGLAND QUARTERLY, April, 1855.

III. *A General View of the History of the Canon of* the NEW TESTAMENT during the First Four Centuries.

By BROOKE FOSS WESTCOTT, M.A., Assistant Master of Harrow School, formerly Fellow of Trinity College, Cambridge. Crown 8vo. cloth, 12*s.* 6*d.*

"*The author is one of those who are teaching us that it is possible to rifle the store-houses of German theology, without bearing away the taint of their atmosphere; and to recognise the value of their accumulated treasures, and even track the vagaries of their theoretic ingenuity, without abandoning in the pursuit the clear sight and sound feeling of English common sense......It is by far the best and most complete book of the kind; and we should be glad to see it well placed on the lists of our examining chaplains.*"—GUARDIAN, Oct. 3, 1855.

"*Learned, dispassionate, discriminating, worthy of his subject and the present state of Christian Literature in relation to it.*"—BRITISH QUARTERLY, Oct. 1855.

Theological Manuals.

IN THE PRESS.

A History of the Christian Church during the Reformation. By CHARLES HARDWICK, M.A., Fellow of St Catharine's Hall, Cambridge, Divinity Lecturer of King's College, and Christian Advocate in the University.
In February.

THE FOLLOWING WORKS OF THE SERIES
ARE IN PREPARATION.

An Introduction to the Study of the Old Testament, with an Outline of Scripture History.

Notes, Critical and Explanatory, on the Hebrew Text of the Prophet ISAIAH.

An Introduction to the Study of the Gospels.

——————————— *Epistles.*

Notes, Critical and Explanatory, on the Greek Text of the FOUR GOSPELS AND THE ACTS OF THE APOSTLES.

Notes, Critical and Explanatory, on the Greek Text of the CANONICAL EPISTLES AND THE APOCALYPSE.

A History of the Christian Church during THE FIRST SIX CENTURIES.

A History of the Christian Church from the Beginning of the XVIIth CENTURY TO THE PRESENT TIME.

An Historical Exposition of the Apostles', Nicene, and *Athanasian* CREEDS.

An Exposition of the Articles of the Church of England.

Others are in progress, and will be announced in due time.

LIST OF BOOKS

PUBLISHED BY

MACMILLAN AND CO.

Cambridge,

AND 23, HENRIETTA STREET, COVENT GARDEN, LONDON.

ABDY.—A Historical Sketch of Civil Procedure among the

Romans. By J. T. ABDY, LL.D., Regius Professor of Civil Law in the University of Cambridge. Crown 8vo. cloth, *4s. 6d.*

ÆSCHYLI Eumenides.

The Greek Text with English Notes, and an Introduction, containing an Analysis of Müller's Dissertations. By BERNARD DRAKE, M.A., late Fellow of King's College, Cambridge. 8vo. cloth, *7s. 6d.*

ADAMS.—The Twelve Foundations and other Poems.

By H. C. ADAMS, M.A., Author of "Sivan the Sleeper," &c. Royal 16mo. cloth, *5s.*

AGNES HOPETOUN'S SCHOOLS AND HOLIDAYS.

The Experiences of a Little Girl. A Story for Girls. By Mrs. OLIPHANT, Author of "Margaret Maitland." Royal 16mo. *6s.*

AIRY.—Mathematical Tracts on the Lunar and Planetary

Theories. The Figure of the Earth, Precession and Nutation, the Calculus of Variations, and the Undulatory Theory of Optics. By G. B. AIRY, M.A., Astronomer Royal. **Fourth Edition,** revised and improved. 8vo. cloth, *15s.*

ARISTOTLE on the Vital Principle.

Translated, with Notes. By CHARLES COLLIER, M.D., F.R.S., Fellow of the Royal College of Physicians. Crown 8vo. cloth, *8s. 6d.*

BEAMONT.—Catherine, the Egyptian Slave in 1852.

A Tale of Eastern Life. By W. J. BEAMONT, M.A., Fellow of Trinity College, Cambridge, late Principal of the English College, Jerusalem. Fcap. 8vo. *5s. 6d.*

BEASLEY.—An Elementary Treatise on Plane Trigonometry;

with a numerous Collection of Examples. By R. D. BEASLEY, M.A., Fellow of St. John's College, Cambridge, Head-Master of Grantham Grammar School. Crown 8vo. cloth, *3s. 6d.*

BIRKS.—The Difficulties of Belief in connexion with the

Creation and the Fall. By THOMAS RAWSON BIRKS, M.A., Rector of Kelshall, and EXAMINING CHAPLAIN TO THE LORD BISHOP OF CARLISLE; Author of "The Life of the Rev. E. Bickersteth." Crown 8vo. cloth, *4s. 6d.*

20. 10. 59.
3,000. cr.

BLAKESLEY.—Four Months in Algeria.
With a Visit to Carthage. With Maps and Illustrations after Photographs. By J. W. BLAKESLEY, B.D., Vicar of Ware, and sometime Fellow and Tutor of Trinity College, Cambridge. 8vo. cloth, 14*s*.

BOLTON.—The Evidences of Christianity.
By W. J. BOLTON, of Caius College, Cambridge. 8vo. 6*s*.

BOOLE.—The Mathematical Analysis of Logic.
By GEORGE BOOLE, F.R.S. Professor of Mathematics, in the Queen's University, Ireland. 8vo. sewed, 5*s*.

BOOLE.—A Treatise on Differential Equations.
Crown 8vo. cloth, 14*s*.

BOWEN.—The Force of Habit considered as an Argument to prove the Moral Government of Man by God. By E. E. BOWEN, B.A., Scholar of Trinity College, Cambridge. Crown 8vo. 1*s*.

BRAVE WORDS for BRAVE SOLDIERS and SAILORS.
Tenth Thousand. 16mo. sewed, 2*d*.; or 10*s*. per 100.

BRETT.—Suggestions relative to the Restoration of Suffragan Bishops and Rural Deans. By THOMAS BRETT (A.D. 1711). Edited by JAMES FENDALL, M.A., Procter in Convocation for the Clergy of Ely. Crown 8vo. cloth, 2*s*. 6*d*.

BRIMLEY.—Essays, by the late GEORGE BRIMLEY, M.A.
Edited by W. G. CLARK, M.A., Tutor of Trinity College, and Public Orator in the University of Cambridge. With Portrait. Crown 8vo. cloth, 7*s*. 6*d*.

CONTENTS.—I. Tennyson's Poems.—II. Wordsworth's Poems.—III. Poetry and Criticism.—IV. Angel in the House.—V. Carlyle's Life of Stirling. VI. Esmond.—VII. My Novel.—VIII. Bleak House.—IX. Westward Ho!—X. Wilson's Noctes.—XI. Comte's Positive Philosophy.

BUTLER (Archer).—Sermons, Doctrinal and Practical.
By the Rev. WILLIAM ARCHER BUTLER, M.A. late Professor of Moral Philosophy in the University of Dublin. Edited, with a Memoir of the Author's Life, by the Very Rev. THOMAS WOODWARD, M.A. Dean of Down. With Portrait. **Fifth Edition.** 8vo. cloth, 12*s*.

BUTLER (Archer).—A Second Series of Sermons.
Edited by J. A. JEREMIE, D.D. Regius Professor of Divinity in the University of Cambridge. **Third Edition.** 8vo. cloth, 10*s*. 6*d*.

BUTLER (Archer).—History of Ancient Philosophy.
A Series of Lectures. Edited by WILLIAM HEPWORTH THOMPSON, M.A. Regius Professor of Greek in the University of Cambridge. 2 vols. 8vo. cloth, 1*l*. 5*s*.

BUTLER (Archer).—Letters on Romanism, in Reply to Mr. NEWMAN'S Essay on Development. Edited by the Very Rev. T. WOODWARD, Dean of Down. **Second Edition,** revised by the Ven. Archdeacon HARDWICK. 8vo. cloth, 10*s*. 6*d*.

CAMBRIDGE.—Cambridge Theological Papers. Comprising those given at the Voluntary Theological and Crosse Scholarship Examinations. Edited, with References and Indices, by A. P. MOOR, M.A. of Trinity College, Cambridge, and Sub-warden of St. Augustine's College, Canterbury. 8vo. cloth, 7*s*. 6*d*.

CAMBRIDGE SENATE-HOUSE PROBLEMS and RIDERS, with SOLUTIONS:—

1848—1851.—Problems.	By N. M. FERRERS, M.A. and J. S. JACKSON, M.A. of Caius College. 15*s*. 6*d*.
1848—1851.—Riders.	By F. J. JAMESON, M.A. of Caius College. 7*s*. 6*d*.
1854—Problems and Riders.	By W. WALTON, M.A. of Trinity College, and C. F. MACKENZIE, M.A. of Caius College. 10*s*. 6*d*.
1857—Problems and Riders.	By W. M. CAMPION, M.A. of Queen's College, and W. WALTON, M.A. of Trinity College. 8*s*. 6*d*.

CAMBRIDGE ENGLISH PRIZE POEMS, which have obtained the Chancellor's Gold Medal from the institution of the Prize to 1858. Crown 8vo. cloth, 7*s*. 6*d*.

CAMBRIDGE.—Cambridge and Dublin Mathematical Journal.
The Complete Work, in Nine Vols. 8vo. cloth, 7*l*. 4*s*.
ONLY A FEW COPIES OF THE COMPLETE WORK REMAIN ON HAND.

CAMPBELL.—The Nature of the Atonement and its Relation to Remission of Sins and Eternal Life. By JOHN M'LEOD CAMPBELL, formerly Minister of Row. 8vo. cloth, 10*s*. 6*d*.

CICERO.—Old Age and Friendship.
Translated into English. Two Parts. 12mo. sewed, 2*s*. 6*d*. each.

COLENSO.—The Colony of Natal. A Journal of Ten Weeks' Tour of Visitation among the Colonists and Zulu Kafirs of Natal. By the Right Rev. JOHN WILLIAM COLENSO, D.D. Lord Bishop of Natal, with a Map and Illustrations. Fcap. 8vo. cloth, 5*s*.

COLENSO.—Village Sermons.
Second Edition. Fcap. 8vo. cloth, 2*s*. 6*d*.

COLENSO.—Four Sermons on Ordination, and on Missions
18mo. sewed, 1*s*.

COLENSO.—Companion to the Holy Communion, containing the Service, and Select Readings from the writings of Mr. MAURICE. Edited by the Lord Bishop of Natal. *Fine Edition*, rubricated and bound in morocco, antique style, 6*s*.; or in cloth, 2*s*. 6*d*. *Common Paper*, limp cloth, 1*s*.

COTTON.—Sermons and Addresses delivered in Marlborough College during Six Years by GEORGE EDWARD LYNCH COTTON, D.D., Lord Bishop of Calcutta, and Metropolitan of India. Crown 8vo. cloth, 10*s*. 6*d*.

COTTON.—Sermons: chiefly connected with Public Events of 1854. Fcap. 8vo. cloth, 3*s*.

CROSSE.—An Analysis of Paley's Evidences.
By C. H. CROSSE, M.A. of Caius College, Cambridge. 12mo. boards, 3*s*. 6*d*.

DAVIES.—St. Paul and Modern Thought:
Remarks on some of the Views advanced in Professor Jowett's Commentary on St. Paul. By Rev. J. LL. DAVIES, M.A. Fellow of Trinity College, Cambridge, and Rector of Christ Church, Marylebone. 8vo. sewed, 2*s*. 6*d*.

DAYS OF OLD: Stories from Old English History of the Druids, the Anglo-Saxons, and the Crusades. By the Author of "Ruth and her Friends." Imp. 16mo. cloth, 5*s.*

DEMOSTHENES.—Demosthenes de Corona. The Greek Text, with English Notes. By BERNARD DRAKE, M.A. late Fellow of King's Coll. Cambridge, Editor and Translator of the "Eumenides of Æschylus." Crown 8vo. cloth, 5*s.*

DEMOSTHENES.—Demosthenes on the Crown. Translated by J. P. NORRIS, M.A. Fellow of Trinity College, Cambridge, and one of Her Majesty's Inspectors of Schools. Crown 8vo. cloth, 3*s.*

DREW.—A Geometrical Treatise on Conic Sections, with Copious Examples from the Cambridge Senate House Papers. By W. H. DREW, M.A. of St. John's College, Cambridge, Second Master of Blackheath Proprietary School. Crown 8vo. cloth, 4*s.* 6*d.*

EVANS.—Sonnets on the Death of Wellington. By SEBASTIAN EVANS, of Emmanuel College, Cambridge. 8vo. sewed, 1*s.*

FARRAR.—The Influence of Classical Studies on English Literature in the Reigns of Elizabeth and James I. By F. W. FARRAR, Fellow of Trinity College, Cambridge. 8vo. sewed, 1*s.* 6*d.*

FARRAR.—Lyrics of Life. By FREDERIC W. FARRAR, Fellow of Trinity College, Cambridge. Author of "Eric," &c. Fcap. 8vo. cloth, 4*s.* 6*d.*

FORD.—Steps to the Sanctuary; or, the Order for Morning Prayer, set forth and explained in Verse. By JAMES FORD, M.A., Prebendary of Exeter Cathedral. Crown 8vo. cloth, 2*s.* 6*d.*

FROST.—The First Three Sections of Newton's Principia. With Notes and Problems in illustration of the subject. By PERCIVAL FROST, M.A. late Fellow of St. John's College, Cambridge, and Mathematical Lecturer of Jesus College. Crown 8vo. cloth, 10*s.* 6*d.*

GILL.—The Anniversaries. Poems in Commemoration of Great Men and Great Events. By T. H. GILL. Fcap. 8vo. cloth, 5*s.*

GODFRAY.—An Elementary Treatise on the Lunar Theory. With a brief Sketch of the History of the Problem up to the time of Newton. By HUGH GODFRAY, M.A. of St. John's College, Esquire Bedell in the University of Cambridge. 8vo. cloth, 5*s.* 6*d.*

GRANT.—Plane Astronomy. Including Explanations of Celestial Phenomena, and Descriptions of Astronomical Instruments. By A. R. GRANT, M.A., one of Her Majesty's Inspectors of Schools, late Fellow of Trinity College, Cambridge. 8vo. boards, 6*s.*

GRANT.—The Education of the Poor a Blessing not an Injury to the Nation. By the Rev. A. R. GRANT, M.A. Crown 8vo. 6*d.*

GRIFFIN.—The Theory of Double Refraction. By W. N. GRIFFIN, B.D., late Fellow of St. John's College, Cambridge. 8vo. 2*s.*

HALE.—The Transportation Question; or, Why Western Australia should be made a Reformatory Colony instead of a Penal Settlement. By the Right Rev. MATTHEW B. HALE, D.D. Lord Bishop of Perth, W.A. Crown 8vo. sewed 2*s.* 6*d.*

HALLIFAX.—Analysis of the Civil Law, being a comparison of the Roman and English Law. By BISHOP HALLIFAX, with Additions, by J. W. GELDART, LL.D. of Trinity Hall, Cambridge. 8vo. boards, 8*s*. 6*d*.

HAMILTON.—On Truth and Error: Thoughts, in Prose and Verse, on the Principles of Truth, and the Causes and Effects of Error. By JOHN HAMILTON, Esq. (of St. Ernan's), M.A. St. John's College, Cambridge. Crown 8vo. cloth, 10*s*. 6*d*.

HARE.—Charges delivered during the Years 1840 to 1854. With Notes on the Principal Events affecting the Church during that period. By JULIUS CHARLES HARE, M.A. sometime Archdeacon of Lewes, and Chaplain in Ordinary to the Queen. With an Introduction, explanatory of his position in the Church with reference to the parties which divide it, 3 vols. 8vo. cloth, 1*l*. 11*s*. 6*d*.

HARE.—Miscellaneous Pamphlets on some of the Leading Questions agitated in the Church during the Years 1845—51. 8vo. cloth, 12*s*.

HARE.—The Victory of Faith.
Second Edition. 8vo. cloth, 5*s*.

HARE.—The Mission of the Comforter.
Second Edition. With Notes. 8vo. cloth, 12*s*.

HARE.—Vindication of Luther from his English Assailants.
Second Edition. 8vo. cloth, 7*s*.

HARE.—Parish Sermons.
Second Series. 8vo. cloth, 12*s*.

HARE.—Sermons Preacht on Particular Occasions.
8vo. cloth, 12*s*.

*** The two following Books are included in the Three Volumes of Charges, and may still be had separately.

HARE.—The Contest with Rome.
With Notes, especially in answer to Dr. Newman's Lectures on Present Position of Catholics. **Second Edition.** 8vo. cloth, 10*s*. 6*d*.

HARE.—Charges delivered in the Years 1843, 1845, 1846.
Never before published. With an Introduction, explanatory of his position in the Church with reference to the parties which divide it. 6*s*. 6*d*.

HARE.—Portions of the Psalms in English Verse.
Selected for Public Worship. 18mo. cloth, 2*s*. 6*d*.

HARE.—Two Sermons preached in Herstmonceux Church, on Septuagesima Sunday, 1855, being the Sunday after the Funeral of the Venerable Archdeacon Hare. By the Rev. H. VENN ELLIOTT, Perpetual Curate of St. Mary's, Brighton, late Fellow of Trinity College, Cambridge, and the Rev. J. N. SIMPKINSON, Rector of Brington, Northampton, formerly Curate of Herstmonceux. 8vo. 1*s*. 6*d*.

HARDWICK.—Christ and other Masters.
A Historical Inquiry into some of the chief Parallelisms and Contrasts between Christianity and the Religious Systems of the Ancient World. With special reference to prevailing Difficulties and Objections. By the Ven. ARCHDEACON HARDWICK. Part I. INTRODUCTION. PART II. THE RELIGIONS OF INDIA. Part III. THE RELIGIONS OF CHINA, AMERICA, AND OCEANICA. Part IV. RELIGIONS OF EGYPT AND MEDO-PERSIA. 8vo. cloth. 7*s*. 6*d*. each part.

HARDWICK.—A History of the Christian Church, during the Middle Ages and the Reformation. (A.D. 590–1600.)
By Archdeacon Hardwick. Two vols. crown 8vo. cloth, 21*s*.
Vol. I. History from Gregory the Great to the Excommunication of Luther. With Maps.
Vol. II. History of the Reformation of the Church.
Each volume may be had separately. Price 10*s*. 6*d*.
*** These Volumes form part of the Series of Theological Manuals.

HARDWICK.—Twenty Sermons for Town Congregations.
Crown 8vo. cloth, 6*s*. 6*d*.

HAYNES.—Outlines of Equity. By FREEMAN OLIVER HAYNES, Barrister-at-Law, late Fellow of Caius College, Cambridge. Crown 8vo. cloth, 10*s*.

HEDDERWICK.—Lays of Middle Age, and other Poems.
By JAMES HEDDERWICK. Fcp. 8vo. 5*s*.

HEMMING.—An Elementary Treatise on the Differential and Integral Calculus. By G. W. HEMMING, M.A. Fellow of St. John's College, Cambridge. **Second Edition.** 8vo. cloth, 9*s*.

HERVEY.—The Genealogies of our Lord and Saviour Jesus Christ, as contained in the Gospels of St. Matthew and St. Luke, reconciled with each other and with the Genealogy of the House of David, from Adam to the close of the Canon of the Old Testament, and shown to be in harmony with the true Chronology of the Times. By Lord ARTHUR HERVEY, M.A. Rector of Ickworth. 8vo. cloth, 10*s*. 6*d*.

HERVEY.—The Inspiration of Holy Scripture.
Five Sermons preached before the University of Cambridge. 8vo. cloth, 3*s*. 6*d*.

HOWARD.—The Pentateuch; or, the Five Books of Moses.
Translated into English from the Version of the LXX. With Notes on its Omissions and Insertions, and also on the Passages in which it differs from the Authorised Version. By the Hon. HENRY HOWARD, D.D. Dean of Lichfield. Crown 8vo. cloth. GENESIS, 1 vol. 8*s*. 6*d*.; EXODUS AND LEVITICUS, 1 vol. 10*s*. 6*d*.; NUMBERS AND DEUTERONOMY, 1 vol. 10*s*. 6*d*.

HUMPHRY.—The Human Skeleton (including the Joints).
By GEORGE MURRAY HUMPHRY, M.D. Cantab. F.R.C.S., Surgeon to Addenbrooke's Hospital, Lecturer on Surgery and Anatomy in the Cambridge University Medical School. With Two Hundred and Sixty Illustrations drawn from Nature. Medium 8vo. cloth, 1*l*. 8*s*.

INGLEBY.—Outlines of Theoretical Logic.
Founded on the New Analytic of SIR WILLIAM HAMILTON. Designed for a Text-book in Schools and Colleges. By C. MANSFIELD INGLEBY, M.A. of Trinity College, Cambridge. In fcap. 8vo. cloth, 3*s*. 6*d*.

JAMESON.—Analogy between the Miracles and Doctrines of Scripture. By F. J. JAMESON, M.A., Fellow of St. Catharine's College, Cambridge. Fcap. 8vo. cloth, 2*s*.

JAMESON.—Brotherly Counsels to Students. Four Sermons preached in the Chapel of St. Catharine's College, Cambridge. By F. J. JAMESON, M.A. Fcap. 8vo. limp cloth, red edges, 1*s*. 6*d*.

JEWELL.—An Apology of the Church of England, and an Epistle to Seignior Scipio concerning the Council of Trent, translated from the original Latin, and illustrated with Notes, chiefly drawn from the Author's "Defence of the Apology." By A. T. RUSSELL. Fcp. 8vo. bds. 5*s*.

JUSTIN MARTYR.—S. Justini Philosophi et Martyris Apologia Prima. Edited, with a corrected Text, and English Introduction with explanatory Notes, by W. TROLLOPE, M.A. Pembroke College, Cambridge. 8vo. bds. 7*s*. 6*d*.

JUVENAL.—Juvenal, for Schools. With English Notes. By J. E. B. MAYOR, M.A. Fellow and Assistant Tutor of St. John's College, Cambridge. Crown 8vo. cloth, 10*s*. 6*d*.

KENNEDY.—The Influence of Christianity on International Law. By C. M. KENNEDY, of Caius College, Cambridge. Crown 8vo. cloth, 4*s*.

KINGSLEY.—Two Years Ago. By CHARLES KINGSLEY, F.S.A. Rector of Eversley, and Chaplain in Ordinary to the Queen. **Second Edition.** 3 vols. crown 8vo. cloth, 1*l*. 11*s*. 6*d*.

KINGSLEY.—"Westward Ho!" or, the Voyages and Adventures of Sir Amyas Leigh, Knight of Burrough, in the County of Devon, in the Reign of Her Most Glorious Majesty Queen Elizabeth. **New and Cheaper Edition.** Crown 8vo. cloth, 6*s*.

KINGSLEY.—Glaucus; or, the Wonders of the Shore. **New and Illustrated Edition,** corrected and enlarged. Containing beautifully Coloured Illustrations of the Objects mentioned in the Work. Elegantly bound in cloth, with gilt leaves. 7*s*. 6*d*.

KINGSLEY.—The Heroes: or, Greek Fairy Tales for my Children. With Eight Illustrations, Engraved by WHYMPER. **New Edition,** printed on toned paper, and elegantly bound in cloth, with gilt leaves, Imp. 16mo. 5*s*.

KINGSLEY.—Alexandria and Her Schools: being Four Lectures delivered at the Philosophical Institution, Edinburgh. With a Preface. Crown 8vo. cloth, 5*s*.

KINGSLEY.—Phaethon; or Loose Thoughts for Loose Thinkers. **Third Edition.** Crown 8vo. boards, 2*s*.

KINGSLEY.—The Recollections of Geoffry Hamlyn. By HENRY KINGSLEY, Esq. 3 Vols. 1*l*. 11*s*. 6*d*.

LATHAM.—The Construction of Wrought-Iron Bridges, embracing the Practical Application of the Principles of Mechanics to Wrought-Iron Girder Work. By J. H. LATHAM, Esq. Civil Engineer. 8vo. cloth. With numerous detail Plates. 15*s*.

LECTURES TO LADIES ON PRACTICAL SUBJECTS.

Third Edition, revised. Crown 8vo. cloth, *7s. 6d.* By Reverends F. D. MAURICE, CHARLES KINGSLEY, J. LL. DAVIES, ARCHDEACON ALLEN, DEAN TRENCH, PROFESSOR BREWER, DR. GEORGE JOHNSON, DR. SIEVEKING, DR. CHAMBERS, F. J. STEPHEN, Esq., and TOM TAYLOR, Esq.

CONTENTS:—Plan of Female Colleges—The College and the Hospital—The Country Parish — Over Work and Anxiety — Dispensaries — District Visiting—Influence of Occupation on Health—Law as it affects the Poor—Everyday Work of Ladies—Teaching by Words—Sanitary Law—Workhouse Visiting.

LUDLOW.—British India; its Races, and its History,

down to 1857. By JOHN MALCOLM LUDLOW, Barrister-at-Law. 2 vols. fcap. 8vo. cloth, *9s.*

LUSHINGTON.—La Nation Boutiquière: and other Poems,

chiefly Political. With a Preface. By the late HENRY LUSHINGTON, Chief Secretary to the Government of Malta. **Points of War.** By FRANKLIN LUSHINGTON, Judge in the Supreme Courts of the Ionian Isles. In 1 vol. fcap. 8vo. cloth, *3s.*

LUSHINGTON.—The Italian War 1848-9, and the Last

Italian Poet. By the late HENRY LUSHINGTON, Chief Secretary to the Government of Malta. With a Biographical Preface by G. S. VENABLES. Crown 8vo. cloth, *6s. 6d.*

MACKENZIE.—The Christian Clergy of the first Ten Centuries,

and their Influence on European Civilization. By HENRY MACKENZIE, B.A. Scholar of Trinity College, Cambridge. Crown 8vo. cloth, *6s. 6d.*

MACLEAR.—The Christian Statesman and Our Indian Empire.

By G. F. MACLEAR, B.A. Curate of St. Barnabas, Kennington. **Second Edition.** Crown 8vo. cloth, *3s. 6d.*

MANSFIELD.—Paraguay, Brazil, and the Plate.

With a Map, and numerous Woodcuts. By CHARLES MANSFIELD, M.A. of Clare College, Cambridge. With a Sketch of his Life. By the Rev. CHARLES KINGSLEY. Crown 8vo. cloth, *12s. 6d.*

M'COY.—Contributions to British Palæontology; or, First Descriptions

of several hundred Fossil Radiata, Articulata, Mollusca, and Pisces from the Tertiary, Cretaceous, Oolitic, and Palæozoic Strata of Great Britain. With numerous Woodcuts. By FREDERICK M'COY, F.G.S., Professor of Natural History in the University of Melbourne. 8vo. cloth, *9s.*

MASSON.—Essays, Biographical and Critical; chiefly on the English Poets.

By DAVID MASSON, M.A. Professor of English Literature in University College, London. 8vo. cloth, *12s. 6d.*

CONTENTS:—I. Shakespeare and Goethe.—II. Milton's Youth.—III. The Three Devils: Luther's, Milton's, and Goethe's.—IV. Dryden, and the Literature of the Restoration.—V. Dean Swift.—VI. Chatterton: a Story of the Year 1770.—VII. Wordsworth.—VIII. Scottish Influence on British Literature.—IX. Theories of Poetry.—X. Prose and Verse De Quincey.

MASSON.—British Novelists and their Styles; being a Critical Sketch of the History of British Prose Fiction. By DAVID MASSON, M.A. Crown 8vo. cloth, *7s. 6d.*

MASSON.—Life of John Milton, narrated in Connexion with the Political, Ecclesiastical, and Literary History of his Time. Vol. I. with Portraits. *18s.*

MAURICE.—Expository Works on the Holy Scriptures. By FREDERICK DENISON MAURICE, M.A., Chaplain of Lincoln's Inn.

I.—The Patriarchs and Lawgivers of the Old Testament. **Second Edition.** Crown 8vo. cloth, *6s.*

This volume contains Discourses on the Pentateuch, Joshua, Judges, and the beginning of the First Book of Samuel.

II.—The Prophets and Kings of the Old Testament. **Second Edition.** Crown 8vo. cloth, *10s. 6d.*

This volume contains Discourses on Samuel I. and II., Kings I. and II., Amos, Joel, Hosea, Isaiah, Micah, Nahum, Habakkuk, Jeremiah, and Ezekiel.

III.—The Unity of the New Testament. 8vo. cloth, *14s.*

This Volume contains Discourses on the Gospels of St. Matthew, St. Mark, and St. Luke; the Acts of the Apostles; the Epistles of St. James, St. Jude, St. Peter, and St. Paul.

IV.—The Gospel of St. John; a Series of Discourses. **Second Edition.** Crown 8vo. cloth, *10s. 6d.*

V.—The Epistles of St. John; a Series of Lectures on Christian Ethics. Crown 8vo. cloth, *7s. 6d.*

MAURICE.—Expository Works on the Prayer-Book.

I.—The Ordinary Services. **Second Edition.** Fcap. 8vo. cloth, *5s. 6d.*

II.—The Church a Family. Twelve Sermons on the Occasional Services. Fcap. 8vo. cloth, *4s. 6d.*

MAURICE.—What is Revelation? A Series of Sermons on the Epiphany; to which are added Letters to a Theological Student on the Bampton Lectures of Mr. MANSEL. Crown 8vo. cloth, *10s. 6d.*

MAURICE.—Lectures on Ecclesiastical History. 8vo. cloth, *10s. 6d.*

MAURICE.—Theological Essays. **Second Edition,** with a new Preface and other additions. Crown 8vo. cloth, *10s. 6d.*

MAURICE.—The Doctrine of Sacrifice deduced from the Scriptures. With a Dedicatory Letter to the Young Men's Christian Association. Crown 8vo. cloth, *7s. 6d.*

MAURICE.—Christmas Day, and other Sermons. 8vo. cloth, *10s. 6d.*

MAURICE.—The Religions of the World, and their Relations to Christianity. **Third Edition.** Fcap. 8vo. cloth, *5s.*

MAURICE.—On the Lord's Prayer.
Third Edition. Fcap. 8vo. cloth, 2*s.* 6*d.*

MAURICE.—On the Sabbath Day: the Character of the Warrior; and on the Interpretation of History. Fcap. 8vo. cloth, 2*s.* 6*d.*

MAURICE.—Learning and Working.—Six Lectures on the Foundation of Colleges for Working Men, delivered in Willis's Rooms, London, in June and July, 1854. Crown 8vo. cloth, 5*s.*

MAURICE.—The Indian Crisis. Five Sermons.
Crown 8vo. cloth, 2*s.* 6*d.*

MAURICE.—Law's Remarks on the Fable of the Bees.
Edited, with an Introduction of Eighty Pages, by FREDERICK DENISON MAURICE, M.A. Chaplain of Lincoln's Inn. Fcp. 8vo. cloth, 4*s.* 6*d.*

MAURICE.—Miscellaneous Pamphlets:—

I.—Eternal Life and Eternal Death.
Crown 8vo. sewed, 1*s.* 6*d.*

II.—Death and Life. A Sermon. In Memoriam C. B. M.
8vo. sewed, 1*s.*

III.—Plan of a Female College for the Help of the Rich and of the Poor. 8vo. 6*d.*

IV.—Administrative Reform.
Crown 8vo. 3*d.*

V.—The Word "Eternal," and the Punishment of the Wicked. **Fifth Thousand.** 8vo. 1*s.*

VI.—The Name "Protestant:" and the English Bishopric at Jerusalem. **Second Edition.** 8vo. 3*s.*

VII.—Right and Wrong Methods of Supporting Protestantism. 8vo. 1*s.*

VIII.—Thoughts on the Oxford Election of 1847.
8vo. 1*s.*

IX.—The Case of Queen's College, London.
8vo. 1*s.* 6*d.*

X.—The Worship of the Church a Witness for the Redemption of the World. 8vo. sewed, 1*s.*

MAYOR.—Cambridge in the Seventeenth Century.
2 vols. fcap. 8vo. cloth, 13*s.*
Vol. I. Lives of Nicholas Ferrar.
Vol. II. Autobiography of Matthew Robinson.
By JOHN E. B. MAYOR, M.A. Fellow and Assistant Tutor of St. John's College, Cambridge.
*** The Autobiography of Matthew Robinson may be had separately, price 5*s.* 6*d.*

MAYOR.—Early Statutes of St. John's College, Cambridge.
Now first edited with Notes. Royal 8vo. 18*s.*
*** The First Part is now ready for delivery.

MAXWELL.—The Stability of the Motion of Saturn's Rings. By J. C. MAXWELL, M.A. Professor of Natural Philosophy in the University of Aberdeen. 4to. sewed, *6s.*

MOORE.—A New Proof of the Method of Algebra commonly called "Greatest Common Measure." By B. T. MOORE, B.A., Fellow of Pembroke College, Cambridge. Crown 8vo. *6d.*

MORGAN.—A Collection of Mathematical Problems and Examples. Arranged in the Different Subjects progressively, with Answers to all the Questions. By H. A. MORGAN, M.A., Fellow of Jesus College. Crown 8vo. cloth, *6s. 6d.*

MORSE.—Working for God, and other Practical Sermons. By FRANCIS MORSE, M.A. Incumbent of St. John's, Ladywood, Birmingham. **Second Edition.** Fcap. 8vo. cloth, *5s.*

NAPIER.—Lord Bacon and Sir Walter Raleigh. Critical and Biographical Essays. By MACVEY NAPIER, late Editor of the *Edinburgh Review* and of the *Encyclopædia Britannica.* Post 8vo. cloth, *7s. 6d.*

NORRIS.—Ten School-Room Addresses. Edited by J. P. NORRIS, M.A. Fellow of Trinity College, and one of Her Majesty's Inspectors of Schools. 18mo. sewed, *8d.*

NORWAY AND SWEDEN.—A Long Vacation Ramble in 1856. By X and Y. Crown 8vo. cloth, *6s. 6d.*

OCCASIONAL PAPERS on UNIVERSITY and SCHOOL MATTERS; containing an Account of all recent University Subjects and Changes. Two Parts are now ready, price *1s.* each.

PARKINSON.—A Treatise on Elementary Mechanics. For the Use of the Junior Classes at the University, and the Higher Classes in Schools. With a Collection of Examples. By S. PARKINSON, B.D. Fellow and Assistant Tutor of St. John's College, Cambridge. Crown 8vo. cloth, *9s. 6d.*

PARKINSON.—A Treatise on Optics. Crown 8vo. cloth, *10s. 6d.*

PARMINTER.—Materials for a Grammar of the Modern English Language. Designed as a Text-book of Classical Grammar for the use of Training Colleges, and the Higher Classes of English Schools. By GEORGE HENRY PARMINTER, of Trinity College, Cambridge; Rector of the United Parishes of SS. John and George, Exeter. Fcap. 8vo. cloth, *3s. 6d.*

PEROWNE.—"Al-Adjrumiieh." An Elementary Arabic Grammar. By J. J. S. PEROWNE, B.D. Lecturer in Divinity in King's College, London, and Examining Chaplain to the Lord Bishop of Norwich. 8vo. cloth, *5s.*

PERRY.—Sermons Preached before the University of Cam-bridge. By the Right Rev. CHARLES PERRY, Lord Bishop of Melbourne, formerly Fellow and Tutor of Trinity College, Cambridge. Crown 8vo. cloth, *3s.*

PHEAR.—Elementary Hydrostatics.
By J. B. Phear, M.A. Fellow of Clare College, Cambridge. **Second Edition.** Accompanied by numerous Examples, with the Solutions. Crown 8vo. cloth, 5*s*. 6*d*.

PLAIN RULES ON REGISTRATION OF BIRTHS AND DEATHS. Crown 8vo. sewed, 1*d*.; 9*d*. per dozen; 5*s*. per 100.

PLATO.—The Republic of Plato.
Translated into English, with Notes. By Two Fellows of Trinity College, Cambridge, (J. Ll. Davies M.A., and D. J. Vaughan, M.A.) **Second Edition.** 8vo. cloth, 10*s*. 6*d*.

PRINCIPLES of ETHICS according to the NEW TESTAMENT. Crown 8vo. sewed, 2*s*.

PROCTER.—A History of the Book of Common Prayer: with a Rationale of its Offices. By FRANCIS PROCTER, M.A., Vicar of Witton, Norfolk, and late Fellow of St. Catherine's College. **Fourth Edition,** revised and enlarged. Crown 8vo. cloth, 10*s*. 6*d*.

*** This forms part of the Series of Theological Manuals.

PUCKLE.—An Elementary Treatise on Conic Sections and Algebraic Geometry. With a numerous collection of Easy Examples progressively arranged, especially designed for the use of Schools and Beginners. By G. HALE PUCKLE, M.A., Principal of Windermere College. **Second Edition,** enlarged and improved. Crown 8vo. cloth, 7*s*. 6*d*.

RAMSAY.—The Catechiser's Manual; or, the Church Catechism illustrated and explained, for the use of Clergymen, Schoolmasters, and Teachers. By ARTHUR RAMSAY, M.A. of Trinity College, Cambridge. 18mo. cloth, 3*s*. 6*d*.

REICHEL.—The Lord's Prayer and other Sermons.
By C. P. REICHEL, B.D., Professor of Latin in the Queen's University; Chaplain to his Excellency the Lord-Lieutenant of Ireland; and late Donnellan Lecturer in the University of Dublin. Crown 8vo. cloth, 7*s*. 6*d*.

ROBINSON.—Missions urged upon the State, on Grounds both of Duty and Policy. By C. K. ROBINSON, M.A. Fellow and Assistant Tutor of St. Catherine's College. Fcap. 8vo. cloth, 3*s*.

RUTH AND HER FRIENDS. A Story for Girls.
With a Frontispiece. **Second Edition.** Fcap. 8vo. cloth, 5*s*.

SALLUST.—Sallust for Schools.
With English Notes. **Second Edition.** By CHARLES MERIVALE, B.D.; late Fellow and Tutor of St. John's College, Cambridge, &c., Author of the "History of Rome," &c. Fcap. 8vo. cloth, 4*s*. 6*d*.

"THE JUGURTHA" AND "THE CATILINA" MAY BE HAD SEPARATELY, price 2*s*. 6*d*. EACH IN CLOTH.

SCOURING OF THE WHITE HORSE; or, The Long Vacation Ramble of a London Clerk. By the Author of "Tom Brown's School Days." Illustrated by DOYLE. **Eighth Thousand.** Imp. 16mo. cloth, elegant, 8*s*. 6*d*.

SELWYN.—The Work of Christ in the World.
Sermons preached before the University of Cambridge. By the Right Rev. GEORGE AUGUSTUS SELWYN, D.D. Bishop of New Zealand, formerly Fellow of St John's College. **Third Edition.** Crown 8vo. 2*s*.

SELWYN.—A Verbal Analysis of the Holy Bible.
Intended to facilitate the translation of the Holy Scriptures into Foreign Languages. Compiled for the use of the Melanesian Mission. Small folio, cloth, 14*s*.

SIMPSON.—An Epitome of the History of the Christian Church during the first Three Centuries and during the Reformation. With Examination Papers. By WILLIAM SIMPSON, M.A. **Third Edition.** Fcp. 8vo. cloth, 5*s*.

SMITH.—City Poems.
By ALEXANDER SMITH, Author of "A Life Drama," and other Poems. Fcap. 8vo. cloth. 5*s*.

SMITH.—Arithmetic and Algebra, in their Principles and Application: with numerous systematically arranged Examples, taken from the Cambridge Examination Papers. By BARNARD SMITH, M.A., Fellow of St. Peter's College, Cambridge. **Sixth Edition.** Crown 8vo. cloth, 10*s*. 6*d*.

SMITH.—Arithmetic for the use of Schools.
New Edition. Crown 8vo. cloth, 4*s*. 6*d*.

SMITH.—A Key to the Arithmetic for Schools.
Crown 8vo. cloth, 8*s*. 6*d*.

SMITH.—The Koran in India.
By LUMLEY SMITH, Fellow of Trinity Hall, Cambridge. 8vo. sewed, 2*s*.

SNOWBALL.—The Elements of Plane and Spherical Trigonometry. By J. C. SNOWBALL, M.A. Fellow of St. John's College, Cambridge. **Ninth Edition.** Crown 8vo. cloth, 7*s*. 6*d*.

SNOWBALL.—Introduction to the Elements of Plane Trigo-nometry for the use of Schools. **Second Edition.** 8vo. sewed, 5*s*.

SNOWBALL.—The Cambridge Course of Elementary Mechanics and Hydrostatics. Adapted for the use of Colleges and Schools. With numerous Examples and Problems. **Fourth Edition.** Crown 8vo. cloth, 5*s*.

SPRAY.
Crown 8vo. boards, 3*s*.

SWAINSON.—A Handbook to Butler's Analogy.
By C. A. SWAINSON, M.A. Principal of the Theological College, and Prebendary of Chichester Crown 8vo. sewed, 2*s*.

SWAINSON.—The Creeds of the Church in their Relations to Holy Scripture and the Conscience of the Christian. 8vo. cloth, 9*s*.

CONTENTS:—I. Faith in God.—II. Exercise of our Reason.—III. Origin and Authority of Creeds.—IV. Inductive Proof of the Creeds.—V. Continual Guidance of the Spirit.—VI. Test and Application of Scripture.—VII. Private Judgment.—VIII. Strengthening of the Judgment and the Preparation for Controversy. With an Appendix.

TAIT and STEELE.—A Treatise on Dynamics, with nume-rous Examples. By P. G. TAIT, Fellow of St. Peter's College, Cambridge, and Professor of Mathematics in Queen's College, Belfast, and W. J. STEELE, late Fellow of St. Peter's College. Crown 8vo. cloth, 10*s*. 6*d*.

TAYLOR.—The Restoration of Belief.

By ISAAC TAYLOR, Esq., Author of "The Natural History of Enthusiasm." Crown 8vo. cloth, 8*s*. 6*d*.

THEOLOGICAL Manuals.

CHURCH HISTORY: DURING THE MIDDLE AGES AND THE REFORMATION (A.D. 590—1600). By ARCHDEACON HARDWICK. With Four Maps, 2 vols. Crown 8vo. cloth, price 10*s*. 6*d*. each.

THE COMMON PRAYER: ITS HISTORY AND RATIONALE. By FRANCIS PROCTER. **Fourth Edition.** Crown 8vo. cloth, 10*s*. 6*d*.

HISTORY OF THE CANON OF THE NEW TESTAMENT. By B. F. WESTCOTT. Crown 8vo. cloth, 12*s*. 6*d*.

*** Others are in progress, and will be announced in due time.

THRING.—A Construing Book.

Compiled by the Rev. EDWARD THRING, M.A. Head Master of Uppingham Grammar School, late Fellow of King's College, Cambridge. Fcap. 8vo. cloth, 2*s*. 6*d*.

THRING.—The Elements of Grammar taught in English.

Second Edition. 18mo. bound in cloth, 2*s*.

THRING.—The Child's Grammar.

Being the substance of the above, with Examples for Practice. Adapted for Junior Classes. **A New Edition.** 18mo. limp cloth, 1*s*.

THRING.—Sermons delivered at Uppingham School.

Crown 8vo. cloth, 5*s*.

THRING.—School Songs.

A Collection of Songs for Schools. With the Music arranged for four Voices. Edited by EDWARD THRING, M.A., Head Master of Uppingham School, and H. RICCIUS. Small folio, 7*s*. 6*d*.

THRUPP.—Antient Jerusalem: a New Investigation into the

History, Topography, and Plan of the City, Environs, and Temple. Designed principally to illustrate the records and prophecies of Scripture. With Map and Plans. By JOSEPH FRANCIS THRUPP, M.A. Vicar of Barrington, Cambridge, late Fellow of Trinity College. 8vo. cloth, 15*s*.

THOMPSON.—A Sermon preached after the Funeral of the

late DEAN PEACOCK. By W. H. THOMPSON, M.A., Regius Professor of Greek in the University of Cambridge. 8vo. 1*s*.

THUCYDIDES, BOOK VI. With English Notes, and a Map.

By PERCIVAL FROST, Jun. M.A. late Fellow of St. John's College, Cambridge. 8vo. 7*s*. 6*d*.

TODHUNTER.—A Treatise on the Differential Calculus.

With numerous Examples. By I. TODHUNTER, M.A., Fellow and Assistant Tutor of St. John's College, Cambridge. **Second Edition.** Crown 8vo. cloth, 10*s*. 6*d*.

TODHUNTER.—A Treatise on the Integral Calculus.

With numerous Examples. Crown 8vo. cloth, 10*s*. 6*d*.

TODHUNTER.—A Treatise on Analytical Statics, with numerous Examples. **Second Edition.** Crown 8vo. cloth, 10*s*. 6*d*.

TODHUNTER.—A Treatise on Conic Sections, with numerous Examples. **Second Edition.** Crown 8vo. cloth, 10*s*. 6*d*.

TODHUNTER.—Algebra for the use of Colleges and Schools. Crown 8vo. cloth, 7*s*. 6*d*.

TODHUNTER.—Plane Trigonometry for Colleges and Schools. Crown 8vo. cloth, 5*s*.

TODHUNTER.—A Treatise on Spherical Trigonometry for the Use of Colleges and Schools. Crown 8vo. cloth, 4*s*. 6*d*.

TODHUNTER.—Examples of Analytical Geometry of Three Dimensions. Crown 8vo. cloth, 4*s*.

TOM BROWN'S SCHOOL DAYS.
By AN OLD BOY. **Sixth Edition.** Crown 8vo. cloth, 10*s*. 6*d*.

TRENCH.—Synonyms of the New Testament.
By The Very Rev. RICHARD CHENEVIX TRENCH, D.D. Dean of Westminster. **Fourth Edition.** Fcap. 8vo. cloth, 5*s*.

TRENCH.—Hulsean Lectures for 1845—46.
CONTENTS. 1.—The Fitness of Holy Scripture for unfolding the Spiritual Life of Man. 2.—Christ the Desire of all Nations; or the Unconscious Prophecies of Heathendom. **Fourth Edition.** Foolscap 8vo. cloth, 5*s*.

TRENCH.—Sermons Preached before the University of Cam- bridge. Fcap. 8vo. cloth, 2*s*. 6*d*.

VAUGHAN.—Notes for Lectures on Confirmation. With suitable Prayers. By C. J. VAUGHAN, D.D., Head Master of Harrow School. **Second Edition.** Limp cloth, red edges, 1*s*. 6*d*.

VAUGHAN.—St. Paul's Epistle to the Romans.
The Greek Text, with English Notes. By C. J. VAUGHAN, D.D. 8vo. cloth, 7*s*. 6*d*.

VAUGHAN.—Sermons preached in St. John's Church, Leicester, during the years 1855 and 1856. By DAVID J. VAUGHAN, M.A. Fellow of Trinity College, Cambridge, and Incumbent of St. Mark's, Whitechapel. Crown 8vo. cloth, 5*s*. 6*d*.

VAUGHAN.—Three Sermons on The Atonement. With a Preface. By D. J. Vaughan, M.A. Limp cloth, red edges, 1*s*. 6*d*.

WAGNER.—Memoir of the Rev. George Wagner, late of St. Stephen's, Brighton. By J. N. SIMPKINSON, M.A. Rector of Brington, Northampton. **Second Edition.** Crown 8vo. cloth, 9*s*.

WATERS OF COMFORT.—A Small Volume of Devotional Poetry of a Practical Character. By the Author of "Visiting my Relations." Fcap. 8vo. cloth, 4*s*.

WESTCOTT.—History of the Canon of the New Testament during the First Four Centuries. By BROOKE FOSS WESTCOTT, M.A., Assistant Master of Harrow School; late Fellow of Trinity College, Cambridge. Crown 8vo. cloth, 12*s*. 6*d*.

*** This forms part of the Series of Theological Manuals.

WESTCOTT.—Characteristics of the Gospel Miracles.
Sermons preached before the University of Cambridge. **With Notes.** By B. F. WESTCOTT, M.A., Author of "History of the New Testament Canon." Crown 8vo. cloth, 4*s*. 6*d*.

WHITMORE.—Gilbert Marlowe and Other Poems.
With a Preface by the Author of "Tom Brown's Schooldays." Fcap. 8vo. cloth, 3*s*. 6*d*.

WILSON.—The Five Gateways of Knowledge.
By GEORGE WILSON, M.D., F.R.S.E., Regius Professor of Technology in the University of Edinburgh. **Second Edition.** Fcap. 8vo. cloth, 2*s*. 6*d*. or in Paper Covers, 1*s*.

WILSON.—The Progress of the Telegraph.
Fcap. 8vo. 1*s*.

WILSON.—A Treatise on Dynamics.
By W. P. WILSON, M.A., Fellow of St. John's, Cambridge, and Professor of Mathematics in the University of Melbourne. 8vo. bds. 9*s*. 6*d*.

WORSHIP OF GOD AND FELLOWSHIP AMONG MEN.
A Series of Sermons on Public Worship. Fcap. 8vo. cloth, 3*s*. 6*d*.
By F. D. MAURICE, M.A. T. J. ROWSELL, M.A. J. LL. DAVIES, M.A. and D. J. VAUGHAN, M.A.

WRIGHT.—The Iliad of Homer.
Translated into English Verse by J. C. WRIGHT, M.A. Translator of Dante. Crown 8vo. Books I.—VI. 5*s*.

WRIGHT.—Hellenica; or, a History of Greece in Greek,
as related by Diodorus and Thucydides, being a First Greek Reading Book, with Explanatory Notes, Critical and Historical. By J. WRIGHT, M.A., of Trinity College, Cambridge, and Head-Master of Sutton Coldfield Grammar School. **Second Edition,** WITH A VOCABULARY. 12mo. cloth, 3*s*. 6*d*.

WRIGHT.—A Help to Latin Grammar;
or, the Form and Use of Words in Latin. With Progressive Exercises. Crown 8vo. cloth, 4*s*. 6*d*.

WRIGHT.—The Seven Kings of Rome:
An easy Narrative, abridged from the First Book of Livy by the omission of difficult passages, being a First Latin Reading Book, with Grammatical Notes. Fcap. 8vo. cloth, 3*s*.

WRIGHT.—A Vocabulary and Exercises on the "Seven Kings of Rome." Fcap. 8vo. cloth, 2*s*. 6*d*.

*** The Vocabulary and Exercises may also be had bound up with "The Seven Kings of Rome." Price 5*s*. cloth.

THE JOURNAL OF CLASSICAL AND SACRED PHILOLOGY.

No. XI. 4*s*.

Vols. I. to III. are now ready, 12*s*. 6*d*. each, cloth lettered.

CASES CAN BE HAD FOR BINDING THE VOLUMES.

R. CLAY, PRINTER, BREAD STREET HILL.

Made in the USA
Monee, IL
04 November 2026

67385001R00187